Organic Food and Farming in China

Despite reports of food safety and quality scandals, China has a rapidly expanding organic agriculture and food sector, and there is a revolution in ecological food and ethical eating in China's cities. This book shows how a set of social, economic, cultural, and environmental conditions have converged to shape the development of a "formal" organic sector, created by "top-down" state-developed standards and regulations, and an "informal" organic sector, created by "bottom-up" grassroots struggles for safe, healthy, and sustainable food. This is generating a new civil movement focused on ecological agriculture and quality food.

Organic movements and markets have typically emerged in industrialized food systems that are characterized by private land ownership, declining small farm sectors, consolidated farm to retail chains, predominance of supermarket retail, standards and laws to safeguard food safety, and an active civil society sector. The authors contrast this with the Chinese context, with its unique version of "capitalism with socialist characteristics," collective farmland ownership, and predominance of smallholder agriculture and emerging diverse marketing channels. China's experience also reflects a commitment to domestic food security, evolving food safety legislation, and a civil society with limited autonomy from a semi-authoritarian state that keeps shifting the terrain of what is permitted. The book will be of great interest to advanced students and researchers of agricultural and food systems and policy, as well as rural sociology and Chinese studies.

Steffanie Scott is an associate professor in the Department of Geography and Environmental Management, Faculty of Environment, University of Waterloo, Canada.

Zhenzhong Si is a postdoctoral fellow at the Balsillie School of International Affairs, Waterloo, Canada.

Theresa Schumilas is a postdoctoral fellow and research associate, Department of Geography and Environmental Studies, Wilfrid Laurier University, Canada.

Aijuan Chen is a policy analyst at Agriculture and Agri-Food, Canada.

Earthscan Food and Agriculture

Environmental Justice and Soy Agribusiness
Robert Hafner

Farmers' Cooperatives and Sustainable Food Systems in Europe
Raquel Ajates Gonzalez

The New Peasantries
Rural Development in Times of Globalization
(Second Edition)
Jan Douwe van der Ploeg

Sustainable Intensification of Agriculture
Greening the World's Food Economy
Jules N. Pretty and Zareen Pervez Bharucha

Contested Sustainability Discourses in the Agrifood System
Edited by Douglas H. Constance, Jason Konefal and Maki Hatanaka

The Financialization of Agri-Food Systems
Contested Transformations
Edited by Hilde Bjørkhaug, André Magnan and Geoffrey Lawrence

Redesigning the Global Seed Commons
Law and Policy for Agrobiodiversity and Food Security
Christine Frison

Organic Food and Farming in China
Top-down and Bottom-up Ecological Initiatives
Steffanie Scott, Zhenzhong Si, Theresa Schumilas and Aijuan Chen

For further details please visit the series page on the Routledge website:
www.routledge.com/books/series/ECEFA.

Organic Food and Farming in China

Top-down and Bottom-up Ecological Initiatives

Steffanie Scott, Zhenzhong Si,
Theresa Schumilas, and Aijuan Chen

LONDON AND NEW YORK   from Routledge

First published 2018
by Routledge
2 Park Square, Milton Park, Abingdon, Oxon OX14 4RN

and by Routledge
52 Vanderbilt Avenue, New York, NY 10017, USA

First issued in paperback 2020

Routledge is an imprint of the Taylor & Francis Group, an informa business

British Library Cataloguing-in-Publication Data
A catalogue record for this book is available from the British Library

Library of Congress Cataloging-in-Publication Data
Names: Scott, Steffanie, 1970– author. | Si, Zhenzhong, author. |
Schumilas, Theresa, author. | Chen, Aijuan, author.
Title: Organic food and farming in China : top-down and bottom-up
ecological initiatives / Steffanie Scott, Zhenzhong Si, Theresa
Schumilas and Aijuan Chen.
Description: Milton Park, Abingdon, Oxon ; New York, NY :
Routledge, 2019. |
Series: Earthscan food and agriculture | Includes bibliographical
references and index.
Identifiers: LCCN 2018022754| ISBN 9781138573000 (hbk) |
ISBN 9780203701706 (ebk)
Subjects: LCSH: Organic farming--China.
Classification: LCC S605.5 .S43 2019 | DDC 631.5/840951–dc23
LC record available at https://lccn.loc.gov/2018022754

ISBN 13: 978-0-367-58628-7 (pbk)
ISBN 13: 978-1-138-57300-0 (hbk)

Typeset in Sabon
by Wearset Ltd, Boldon, Tyne and Wear

To the farmers of 40 centuries and the new peasants wearing their hats

Contents

Figures

Tables

Acknowledgements

We are extremely grateful for the time given up by farm operators and staff, farmers' market coordinators, buying club volunteers, government officials, staff of certification agencies, researchers, and others, for sharing their stories and viewpoints with us.

We acknowledge funding from the Social Sciences and Research Council of Canada to carry out the fieldwork on which this research is based. We would like to thank members of the Waterloo Food Issues Group, plus dissertation committee members and colleagues, for constructive input on various drafts and presentations we have made on this research.

Three chapters in this book are based on previously published work, although they have been substantially revised for this book. Chapter 4 is revised and published here with permission of the Lyson Center for Civic Agriculture and Food Systems and the *Journal of Agriculture, Food Systems, and Community Development*, and reworked versions of Chapters 5 and 8 are used with permission from *Agriculture and Human Values* and *Asia Pacific Viewpoint*, respectively.

Acknowledgments

1 Introduction

Steffanie Scott, Zhenzhong Si,
Theresa Schumilas, and Aijuan Chen

The story of this research project

Let us begin this story with our visit in May 2012 to the BioFach China trade fair in Shanghai, the biggest annual organic food trade fair in China. It was held only two months after the enactment of the new and more stringent national organic certification standards. The first floor of the large two-floor venue was filled with booths of organic food companies, local government representatives, farmers' cooperatives, and visitors who were seeking business opportunities. Curiously, on the second floor, there were two separate venues for conferences and seminars. In the large auditorium, the China Organic Congress was being held. CEOs of many well-known organic food companies, along with government officials, gave presentations about the changing organic sector. Meanwhile, just a few steps away from this auditorium, in a much smaller space, a "community-supported agriculture" (CSA) forum was going on. CSA farmers, buying club organizers, and representatives of farmers' markets gave presentations about their initiatives and visions. Unlike the auditorium group, who were discussing the new organic certification standards and marketing, people here were debating about enhancing producer–consumer connections, community building among consumers, and alternative ecological agricultural approaches. We were struck by these two parallel but disconnected "worlds," with very different levels of economic and political power. It was clear to us that, while both of these groups were engaging intensively in ecological agriculture, their understandings, approaches, and goals of ecological agriculture differed significantly. As we continued our exploratory journey in the organic sector, the juxtaposition of these two "worlds" came up again and again. It became a puzzle for us to explain. We asked ourselves how this came about, what were the major features that distinguished these interest groups, and what different socio-economic and regulatory conditions they faced. In this book, we examine the characteristics, conditions, and cases of the two worlds—top-down and bottom-up initiatives in the organic sector.

Backing up further, the original idea for this study was born during a scoping research trip to China by Steffanie Scott in 2009. During this time, she visited several organic farms near Shanghai and Nanjing, met with representatives of the Organic Food Development Centre, and learned about the fast-growing domestic market for organic foods in China. Steffanie applied to the Social Sciences and Humanities Research Council of Canada and was granted funding for the study in 2010. At the time, Steffanie already had one PhD student, Aijuan Chen, who was able to start fieldwork in 2010 to learn more about the role of small farmers in China's ecological agriculture sector. She and Steffanie conducted fieldwork together in spring 2011. A second PhD student, Zhenzhong Si, began his doctoral studies in 2010 and conducted fieldwork (in part with Steffanie) in 2012, and again in 2013. Theresa Schumilas joined the team of doctoral students in early 2011 and carried out fieldwork in 2012 and 2013. Then, in 2014, Steffanie was visiting scholar at the Chinese University of Hong Kong for one semester, and from there gleaned a perspective on organic and ecological farming and organic food consumers in that part of "China."

Our unique team of two Chinese and two Canadians, with diverse backgrounds in rural development issues, the organic sector, and land management, made for deep learning from each other and deep reflection on our fieldwork findings. Along our journeys, we had all sorts of conversations about China, Canada, organic food and farming, agroecology, peasants, farmers, entrepreneurs, back-to-the-land movements, activism, and how these concepts morphed and translated across space and time. In the concluding chapter we provide some more reflections and impressions of fieldwork based on our diverse perspectives.

Our research focus shifted somewhat over time as our understanding deepened. The research began with an exploration of the ecological agriculture sector in China. Over time, we realized that a vibrant informal organic and ecological agriculture sector was emerging outside the formal, certified organic food sector. Some of the early interviewees were involved in CSA farming and ecological farmers' markets. They introduced our research team to more initiators of alternative food networks (AFNs) across the country. We (particularly Zhenzhong Si and Theresa Schumilas) became interested in the characteristics of these AFNs and how they compared with the values of their counterparts in the West.[1] This led us to examine more nuanced tensions within these AFNs and how these tensions shaped their principles and ways of operating. We gradually realized that, in many cases, these various initiatives had strong rural development goals in terms of fostering the well-being of small peasants in rural areas. A new horizon was unveiled when Zhenzhong discovered their connections with the New Rural Reconstruction Movement (NRRM). As his research focus shifted toward this grassroots rural development initiative, he found that it was far more than a campaign of ecological agriculture, although ecological agriculture was a key component of their initiatives. In late 2012, Zhenzhong and Theresa attended the 4th National CSA Symposium, organized by the NRRM team in Beijing, and the International Conference on Rural Reconstruction and Food Sovereignty, held in Chongqing. These two events showcased this multifaceted and vibrant "bottom-up" social movement.

In 2016, Steffanie (with Zhenzhong Si and other collaborators) received another grant from the Social Sciences and Humanities Research Council of Canada, this time focusing more in depth on the dynamics, knowledge networks, and new entrants to the ecological agriculture sector in one region—around Nanjing. This research is still ongoing but we have drawn on a few insights from it, including fieldwork in 2016, to update the findings reported in this book.

Research objectives

When we embarked on this research in earnest in 2010, there was very little published work on China's organic or ecological agriculture sector. Through our exploratory study, we sought to learn about various pieces of this puzzle:

- the evolution (or development path) of China's ecological agriculture sector over time, the types of ownership structures of organic farms, and relationships between government agencies, agribusiness enterprises, private farms of various sizes, and farmers' cooperatives in the sector;
- the type and extent of involvement of small-scale farmers in this sector, in terms of access to land and capital, labour relations, and knowledge networks;
- the engagement and challenges of farmers' cooperatives in this sector;
- the factors shaping the development of AFNs in China, amid unprecedented cultural change, and in a context of a state-driven yet market-oriented economy with limited civil society involvement;
- the adoption within China of alternative values and practices from AFNs internationally; major types AFNs in China, and dimensions of alternativeness, within the socio-political and economic context in China; and
- the co-evolution of, and synergies between, AFNs and rural development initiatives—specifically, the New Rural Reconstruction Movement.

A note on terminology

We would like to share a few reflections on connotations and translations of relevant terminology in Chinese. "Ecological agriculture" (*shengtai nongye*), "sustainable agriculture" (*kechixu nongye*), and "circular agriculture" (*xunhuan nongye*) are the most popular terms used in China in relation to the "greening" of agriculture. Organic agriculture (*youji nongye*) is also an established concept, though it often has the connotation of third-party certification rather than the wider subset of practices that the term "ecological agriculture" can entail (Schumilas 2014). Given that our research started by examining the organic agriculture sector and other ecological agricultural practices, we have chosen to use the term ecological agriculture as our key focus. Ecological agriculture encompasses organic agriculture (certified and otherwise), as well as other efforts toward the "greening" of agriculture, such as certified "green food" in China, and an array of other practices among larger and smaller-scale farms, such as natural and integrated farming, permaculture, circular agriculture (nutrient cycling within a farm), and more. Alternatively, agroecology (*nongye shengtai*) is a term that has gained popularity in some circles internationally among the food sovereignty movement (Holt-Giménez and Altieri 2013) and researchers alike (Gliessman 1990; Altieri 1995). However, this term does not appear to be widely recognized (outside the academic field) within China. Having said that, a group of agroecological scientists (Luo and Gliessman 2016) published an edited volume in English, *Agroecology in China*, in 2016. We expect that this term will gain popularity in the years to come.

Agroecology is usually translated into Chinese as "ecological agriculture" or "eco-agriculture" (Luo 2016; Li 2001, 2003). The first conference on agroecology put on by the China Academy of Sciences was held in 1981 in Nanjing (Luo 2016: 2). In contrast with understandings of agroecology in Latin America and some other places around the world, notions of ecological agriculture and research on ecological agriculture in China tend to be much narrower and shallower (or diluted) in at least two respects. First, they focus mainly on ecological and not human or socioecological dimensions. This is exemplified, for example, in the 1987 book (in Chinese) *Agroecology Engineering in China* (Ma 1987). Second, even within the ecological dimension, many researchers, government officials, and farmers consider a small use of agrochemicals acceptable within their understanding of ecological agriculture. Moreover, the scaling out of organic agriculture is not seen as feasible for China, in the view of most key stakeholders—even those involved in this sector. In the realm of ecological agriculture research, analyses tend to ignore human (i.e. farmer) dimensions or roles, in contrast to a farming systems research approach (Collinson 2000). In addition, they focus on quantitative not qualitative aspects, and tend to overlook small-scale producers and AFNs. Instead, their research always takes an engineering and scientific perspective. Moreover, scientific researchers see themselves as experts, rather than being interested in farmers' local knowledge and farmer participatory research.

In the international agroecology movement, and among many researchers, agroecology is viewed as an approach to promote solidarity among small farmers and to embrace a socio-ecological systems perspective. However, in China, the discussion is more about moderately scaling up family farm operations for greater productivity and economic gains. This focus accords with Chinese government technocratic viewpoints on ecological agriculture that emphasize productivity and technology. There is considerable optimism around "agricultural industrialization" and vertical integration of ecological agriculture among the research community and the government. The current government approaches are also project-based (in silos and specific sites) rather than a systematic, integrated approach. Eco-agritourism is emphasized as the major approach for ecological agriculture development. Despite this clear bias toward government-supported, modernization-oriented development of the ecological agriculture sector, our analysis draws attention to both top-down (or mainstream) ecological agriculture and bottom-up initiatives.

The meanings of "industrial agriculture" are also somewhat different in China from understandings that are common elsewhere. There are two interpretations of industrialization—as an industry versus as an economic sector. The first dimension, known as *gongyehua* in Chinese, refers to the adoption of chemicals, machinery, high-yielding varieties of seeds, and other standardized industrial inputs in farming, and scaling up smallholder agriculture. The other dimension of this, known as *chanyehua* in Chinese, refers to the commercialization and integration of the agriculture sector into the market economy, which aims to increase farmers' income and diversify services on farm, such as by processing food and agritourism. Because of these two different interpretations of the term industrialization in Chinese, English literature that argues against agricultural industrialization may have been misinterpreted as promoting a shift back to subsistence agriculture. Researchers need to be cautious and have a clear definition of industrialization when discussing the implications of industrial agriculture.

There has been a hot debate on the "appropriate" scale of agriculture (*shidu guimo jingying*) in China, a term that often appeared in governmental documents in recent years. Prof Luo Shiming, one of the key advocates and analysts of ecological agriculture developments in China, recommends that ecological agriculture farms be mid- rather than small-scale in order to be economically viable. Luo also argues that we need to think about different types of machinery needed by small ecological agriculture farms. In the meantime, he argues that there needs to be a better regulatory system, compensation for ecosystem services (e.g. saving water, composting on-site) and enforcement (penalties) for farmers overusing agrochemicals or burning rice stalks.

In using the terms top-down and bottom-up in the book's subtitle, we recognize the problems of suggesting a dichotomy in categorizing ecological agricultural initiatives. We are also sensitive to potentially misleading readers into thinking that all top-down initiatives that we discuss are state-led. In practice, most of the ecological agriculture initiatives that we examined are shaped by both state and civil society actors. We chose the term "top-down" because the state plays a stronger role in the creation and development of these initiatives. Moreover, these initiatives tend to be consistent with the state's vision of modern ecological agriculture, in terms of larger-scale, higher-capital investment, specialized production, and deeper integration into mainstream food supply chains. In the end, we opted to use these distinctions as a heuristic to highlight the differing approaches, visions, and values behind the creation of these initiatives.

Research methods

Data collection

For our research, we used multiple qualitative methods to collect and analyze information. Our key method was interviews. Other methods included field visits to farms and alternative food venues, observation of "microblog" and blog posts, and attending CSA symposiums. We also gleaned information from secondary sources including newsletters and informal publications, websites, media coverage, and organic food expos.

Semi-structured interviews were an effective research method for this exploratory type of study as they enabled us to capture opinions of different groups of people and they allowed for open-ended responses and follow-up questions. Moreover, the organic farm sector in China, and particularly AFNs, are nascent initiatives that have not been well documented in existing academic literature. By interviewing people in diverse positions, it is easy to identify not only points of consensus but also disputes and contestations. This is critical for identifying the challenges that confront AFNs and the organic sector in China. Many of the subtleties within these emerging and rapidly evolving initiatives can arise from the interviewees at any time. Some of the interviewees would also help to correct a false perception held by the researcher and disclose misunderstandings that we had not previously identified.

Compared to using a survey questionnaire, interviews enabled us to adjust our questions according to the responses of the respondents and capture critical information. Finally, interviews gave us considerable flexibility to extract information about issues that were of most interest to us. In order to interview people with various backgrounds, we designed various types of questions for a given interviewee. For example, when interviewing a manager of the Beijing Farmer's Market—the most prominent ecological farmers' market in China—we were curious to ask about how the market was initiated, the key rules for selecting vendors, who the vendors are, how it maintains its reputation, the motivations of their customers, their connections with other initiatives and with academics, their perceptions of organic certification, and the core values of the market. But, besides these questions we had prepared, we also learned about the important role of microblogs (known as Weibo in China) in promoting the market, how they had been funding their market, information about

Figure 1.1 Theresa, Steffanie, and Zhenzhong at Green Cow Farm in Beijing.

specific farmers, and the emerging group of "new farmers"—issues we might have missed if we had used a closed-ended survey.

As noted earlier, this study is based on a broad research project about the ecological agriculture sector in China. The research team of three doctoral students (Zhenzhong Si, Theresa Schumilas, and Aijuan Chen) and one professor (Steffanie Scott) collectively conducted 127 interviews in 2011, 2012, and 2013 in 13 provinces and municipalities in China, including Beijing, Liaoning, Shandong, Henan, Anhui, Jiangsu, Shanghai, Zhejiang, Sichuan, Chongqing, Guangxi, Fujian, and Hainan. Our interviewees were key players in the ecological agriculture sector, from diverse positions: employees and owners of organic and green food farms, representatives of organic certification bodies, government agencies, consumer associations, NGOs and community organizers, and researchers (see Table 1.1). Interviews ranged from 30 minutes to five hours. All but five interviews were conducted in Chinese and notes were taken during interviews. Interview notes were later translated and transcribed. We identified most of the interviewees by snowball sampling. The remainder were identified through personal and academic contacts, mass media, online social networks, and national organic conferences and expos.

One online directory that we used early in our study was the China Organic Directory 2009, edited by Organic Services GmbH. It listed organic certification agencies, organic consulting firms, NGOs, and most enterprises and farmers' cooperatives engaged in organic agriculture in China. It helped us to identify potential interviewees and to understand the development and distribution of organic agriculture in China. One

Table 1.1 Number of interviews conducted with different types of interviewees

Type of interviewee	Number of interviews*
Managers and workers on ecological farms	42
Managers of farmers' markets	4
Representatives of buying clubs	3
People renting plots for recreational gardening	5
Governmental officials	20
Researchers	32
Organic certification agencies	11
Directors and employees of NGOs	10
Total	**127**

Note
* Some (repeat) interviews were conducted with the same person.

challenge of using this directory for sampling is that some enterprises and cooperatives listed in the directory had already withdrawn from organic agricultural production by the time we conducted the fieldwork. There is a turnover rate of approximately 30 per cent annually in the sector of organic agriculture. This may happen because organic agriculture is certified annually and some enterprises fail to pass the certification, or some enterprises voluntarily withdraw from organic agriculture for various reasons.[2]

Our second approach for collecting information was field visits to various ecological farms across 13 provinces and municipalities, and to farmers' markets in Beijing and Shanghai. Visiting the ecological farmers' markets, for example, gave us a sense of how vendors promote their products and communicate with customers. By talking with vendors and customers, we collected information about the motivations of customers, the ethical values of vendors, ecological implications of their farming methods, and their perceptions of organic farming and certification. We also participated in a seminar discussion after the market where vendors shared their different perspectives and approaches of how to maintain soil fertility without chemical fertilizers. Through this, we heard about the different farming approaches, farmers' understandings of the principles of organic farming, and their perceptions of organic certification.

Our third approach for collecting information was written text from relevant microblog accounts, blogs, and online forums. "Blog and buzz mining," where internet posts are used as sources of research data, is a relatively new research method in the social sciences (Poynter 2010). Best practices and ethical frameworks are still evolving. Given the conversational nature of blogging, monitoring a community's online exchanges can be similar to monitoring in-person conversations. As such, blogs can be helpful in understanding the beliefs and practices of a particular community. There is an evolving literature around "online activism" specific to

China. The growth of online communities, and in particular the use of micro-blogging has exploded in recent years (Yang 2009). Microblogs have become a significant public space for information flow and exchange since 2011. They have played a critical role in the development of various alternative food initiatives in China.

For their research, Zhenzhong Si and Theresa Schumilas observed various online posts. Zhenzhong's observations of microblogs cover accounts of CSA farms (e.g. Little Donkey Farm, Big Buffalo Farm, Shared Harvest CSA, Emerald Harbor Farm, Tony's Farm), farmers' markets and their organizers and vendors (e.g. Beijing Country Fair Farmer's Market, Beijing Community Farmers' Market, Shanghai Nonghao Farmers' Market, Tianjin Green Farmers' Market, Xi'an Farmers' Market), farmers' market vendors (e.g. Dreamland Farm, Bashangtian Organic Farm, Sunlin Farm, Dandelion Commune, Happy Urban Farmer), buying clubs (e.g. Green League, Shanghai Caituan, Chengdu Green Heartland, Citizen Group of Organic Food Investigation), influential academics and activists in rural development and agriculture (e.g. Li Changping, Jiang Gaoming, Qiu Jiansheng), and alternative food stores (e.g. Jishi, run by Beijing Farmer's Market, Ufood Organic), as well as related organizations and websites (National Urban-Rural Mutual Support CSA Alliance, Hanhaisha, Beijing Organic Assemble, EcoScan, Taobao Ecological Agriculture). When reading these posts, we paid special attention to their opinions and debates about local and seasonal food, trust and community building, self-identity, ecological farming methods, organic certification, healthy eating tips, etc.

In addition to microblogs, Zhenzhong's discourse analysis also included blogs and websites. For example, Shi Yan, the founder of the most influential CSA farm in China—Little Donkey Farm—and several other CSA farms in Beijing, has been an influential figure in the AFNs' community. Her blog posts cover various issues related to AFNs, especially about the values embedded within these networks in the West. Therefore, it provided valuable information to examine food advocacy in China. The New Rural Reconstruction Movement's projects have been widely covered by the mass media. The various websites therefore provided valuable information for understanding how the New Rural Reconstruction Movement addressed its goals with diverse strategies and approaches. Zhenzhong also examined how the mass media described their activities. Critiques of CSA farms and some of their ethical values were also found on the online forum of Emerald Harbor Farm. This provides a contrast, reflecting the contested nature of these nascent alternative food initiatives in China.

During our fieldwork, we learned how extensively AFN participants were using online spaces, so Theresa decided to monitor the Weibo micro-blog posts of eight bloggers for four months. Weibo is a platform that has been in existence since 2009. It is best described as a cross between blogging (as it is understood in North America) and Twitter. The use of Weibo has exploded in the past few years. It had over 100 million users by early

2011 (Yang 2013), and by early 2017 it had 340 million active monthly users, overtaking Twitter (BBC News 2017). Weibo posts include anything from event promotion and distributing information to more political expression. The state censors Weibo for subversive content (Yang 2013) and bloggers typically use pseudonyms to at least partially obscure their identity.

The bloggers that Theresa "followed" were all people involved in the AFNs that we had studied. They included CSA operators, one peasant farmer, buying club volunteers, farmers' market volunteers, and consumers. The blogs were all in Chinese and were translated by a graduate student at the University of Waterloo. Theresa met most of the bloggers while in China. She sought their permission to monitor their blogs, and they consented verbally. Where she did not meet the bloggers personally first, she notified them by email that she was a researcher reading their blogs and that she might quote them in her publications, and asked them to respond if they had any concerns. No one expressed any concern about the blog monitoring.

There is a debate in the research ethics literature as to whether blogs and online discourse should be considered in the public domain, and hence "cited" in the same way as print media references or other "desk research," or whether these postings should be considered a more private form of information sharing and hence subject to the same ethics considerations as other in-person types of qualitative data (Poynter 2010). We treated contributions made in online space the same way as contributions people made in interviews, and assigned numeric codes to mask identities.

A final method used in our research was observation, at various events and forums, including the annual BioFach China organic expo held every May in Shanghai. As we mentioned at the beginning of this chapter, this expo was far more than an exhibition of organic brands and products. It included seminars held by NGOs like Green Ground in Beijing that involved participation from CSA farmers, farmers' markets organizers, and buying clubs organizers. Some of us participated in discussions during these seminars and collected useful information about the challenges and opportunities of AFNs. Another official conference was also held at this expo where large organic food companies, certification agencies, and governmental officials sat together to discuss the development of the organic agriculture sector and policy changes. This forum was in sharp contrast—in terms of values, approaches, and foci—to the AFNs' seminars. Some of our interviews were also conducted at the expo.

Zhenzhong and Theresa also attended two important gatherings held by the New Rural Reconstruction Movement (NRRM) team. One was the 4th National CSA Symposium, at Renmin University in Beijing from November 30 to December 1, 2012. This symposium has been held annually since 2010 by the NRRM. The Rural Reconstruction Center at Renmin University, as the base for the NRRM, facilitated this annual gathering of ecological farmers (including CSA managers), which brought together NGOs, farmers' markets managers and volunteers, buying club organizers, academics, and other coordinators of NRRM initiatives across the country. In the 4th symposium, Zhenzhong and Theresa participated in seminar sessions and roundtable discussions. The information from various presenters, most of whom are organizers and managers of alternative food initiatives, provided a valuable complement to our interviews. The other conference was the International Conference on Sustainability and Rural Reconstruction, held December 8–10, 2012, at Southwest University in Chongqing, China. This was a conference on the alliance between the NRRM in China and rural reconstruction initiatives in many other countries. Academics and activists sat together to address challenges of environmental sustainability, social justice, equity, the economic viability of small-scale farmers, food sovereignty, and food security. The entire conference was pitched with a strong anti-modernity sentiment and alternative development ideas, which is very rare in contemporary China. Attending these events enabled Zhenzhong and Theresa to better understand the NRRM's alternative values and practices. It helped Zhenzhong to learn about the impacts of the food safety crisis on the activities of the NRRM, and the tactics of the NRRM to cope with the state pressure (see Chapter 9).

Data analysis and case study method

In our research, we used various methods to collect qualitative data about ecological agriculture farms, farmers' markets, and buying clubs, and the NRRM as a broad social movement. We opted for a case study method— gathering stories of various ecological agriculture farms from around China—as the best choice for our study. We did this for three main reasons. First, developing case studies through in-depth interviews enabled us to uncover the subtleties existing in the struggles of these farms, and associated alternative food networks, in the Chinese socio-political context. For instance, we would not have been able to illustrate in detail how a farmers' market works in China (i.e. the power struggles and distinct values among different players) without the case analysis of the Beijing Farmers' Market. Second, the case study method helped us to reconsider some preconceptions about the emerging ecological food sector in general. Before doing this fieldwork, some of us assumed that, having been "flooded" with constant food safety scandals, AFNs in China were merely a response to food safety challenges. Without in-depth case studies, we would not have been able to perceive and document the strong ecological and social concerns of CSA farmers and farmers' market managers. This led us to appreciate the differences in values between AFN initiators and their customers. Third, since AFN studies in the West have already developed a set of theories to explain this phenomenon, it was important to examine whether these understandings of AFNs based on Western cases would apply to Chinese AFNs. Case studies enable all these explorations.

In addition to using case studies as a basic research design method, we also used discourse analysis to interrogate the perspectives and value systems of the initiatives that we profiled. When conducting discourse analysis, a key requirement is to interrogate the social context and the power structure within which the discourse is produced (i.e. the "social production" of the text). Discourse analysis fits with our research objectives for several reasons. First, a large proportion of the data we collected was from semi-structured interviews, and interview transcripts reflect nuanced perspectives of interviewees. Therefore, the selection of words, the tone of expressions, and the rationales in the conversation were all important information. When reading the interview transcripts, we put ourselves in the position of the interviewee to understand why certain words were used and why the narratives unfolded in certain ways. Moreover, since for part of our research we sought to understand how the NRRM copes with pressures from the state, we could examine the remarks of the key leader of the NRRM to reveal the political context that shaped how the remarks were phrased and how the NRRM maintained its non-confrontational stance toward the state. The interrogation of the Beijing Farmer's Market also involved an analysis of the WeChat (social media) posts by market staff and the advertisements designed by the market vendors.

Organization of the book

This book is organized as follows. In Chapter 2, we outline some of the context in which China's organic sector development has taken place: the industrialization of agriculture in China, consolidation of food production and processing, supermarketization of food retailing, and changing patterns of food consumption and food safety scandals. We also explain China's food policy regime, which revolves around national food security. This is the context in which the ecological agriculture sector and AFNs in China are emerging and evolving. We characterize the array of food policies in China as a system centring on national food security or, more explicitly, sufficient grain productivity. This policy regime dominated by food security is reflected in various policies revolving around farmland preservation, food reserve and circulation governance, and agricultural policy support. It is also increasingly complemented by concerns of food safety and environmental sustainability of food production. This characterization of China's food policy regime is important for understanding the state's priority of modernization of agriculture. It also provides the context for understanding many of the state impediments and initiatives in ecological agriculture development in China.

Chapter 3 and Chapter 4 elaborate on various agribusiness models in the ecological agriculture sector and the critical roles of the Chinese state in supporting their development. Given the strong state interventions, these business ventures, in many instances, reflect what we call "top-down initiatives" in ecological and organic agriculture. Although not all agribusinesses in the sector receive government support, in many ways their development has been interwoven with broad (state) agendas of agricultural modernization, which strongly support an industrialization and scaling-up of the agriculture sector. These market-driven initiatives, despite their positive environmental implications, have been fostering neo-liberal logics in the agriculture sector and curtailing the already limited roles of small farmers. The role of consumers in these developments is confined to exercising their purchasing power to support the expansion of ecological agribusiness. The direct communication between consumers and these agribusinesses is rather limited owing to the long modern food supply chain through which their food is normally channelled. The creation and maintaining of consumer trust is relies heavily on third-party assurance schemes, such as food quality inspection and monitoring systems or certifications. In this sense, trust is a critical issue for these producers, as we explain later.

In the following chapters, we shift our focus to the recently emerged "bottom-up initiatives" in addressing demands for ecologically produced food. Many of these initiatives happen to relate to a concept widely discussed in the West: alternative food networks (AFNs). In contrast to "top-down initiatives," these AFNs reflect a strong grassroots nature in the sense that their establishment, operations, values, and community organizing strategies all reflect the initiative of producers and consumers rather than the market per se. The nascent and grassroots features of AFNs in China make them a fascinating topic. In Chapters 5, 6, and 7, we strive to understand the typology, alternative characteristics, socio-economic embeddedness, and contestations, as well as the economic, ecological, and interpersonal relations within these initiatives. Several case studies of AFNs in China, particularly the Beijing Farmers' Market, allow us to unveil these various themes.

Chapter 5 depicts the new forms of producer–consumer provisioning networks that are emerging in peri-urban China. We explore the characteristics of AFNs in China and how they differ from AFNs in the West. In Chapter 6, we delve further into the "alternativeness" of these networks along economic, ecological, interpersonal, and political lines. The transformative potential of AFNs, including farmers' markets, to challenge mainstream industrial food supply chains, has been a common theme among food studies scholars. Chapter 7 takes this up by examining various tensions and conflicts that may impinge on the "alternativeness" of one AFN, the most prominent ecological farmers' market in China—the Beijing Farmer's Market. We analyze the contested space of this farmers' market that is created by the interactions among key actors including market managers, vendors, customers, and the state. Chapter 8 then explains how AFNs are moving beyond market-based activities (such as CSAs, farmers' markets, buying clubs) and deploying grassroots community organizing strategies amid the specific socio-political context of contemporary China. Chapter 9 examines the grassroots New Rural Reconstruction Movement (NRRM), the impacts of the food safety crisis on the activities of the NRRM, and the tactics of the NRRM to cope with the state pressure. Through this collection of chapters, we reveal that the alternativeness of AFNs in China is uneven and varies among different elements of alternativeness. In Chapter 10 we draw together some conclusions about the greening of food systems, with Chinese characteristics.

Notes

1 In writing this book, we debated whether to use the more "politically correct" terms of Global North and Global South (also more commonly used by social movements) in place of "the West" and "developing countries." In the end, we opted for the latter given that they are probably more familiar to our readers, particularly readers in China.
2 Interview with the representative of Organic Services GmbH in China, June 7, 2010, in Nanjing, Jiangsu province.

References

Altieri, MA 1995, *Agroecology: The Science of Sustainable Agriculture*. Intermediate Technology Publications, London.

BBC News 2017, "Twitter user numbers overtaken by China's Sina Weibo," May 17. Accessed at www.bbc.com/news/technology-39947442. Accessed July 17, 2018.

Collinson, MP (ed.) 2000, *A History of Farming Systems Research*. CABI and FAO, Wallingford, UK, and Rome.

Gliessman, SR 1990, *Agroecology: Researching the Ecological Basis for Sustainable Agriculture*. Springer, New York.

Holt-Giménez, E and Altieri, MA 2013, "Agroecology, food sovereignty, and the new green revolution," *Agroecology and Sustainable Food Systems*, vol. 37, no. 1, pp. 90–102.

Li, W (ed.) 2001, *Agro-ecological Farming Systems in China*. Parthenon, New York.

Li, W 2003, *Ecoagriculture: Theory and Practice of Sustainable Agriculture in China*. Chemical Engineering Press, Beijing (in Chinese).

Luo, S 2016, "Agroecology development in China—an overview." In: Luo, S and Gliessman, SR (eds.) 2016, *Agroecology in China: Science, Practice, and Sustainable Management*. CRC Press, Boca Raton, FL, pp. 1–35.

Luo, S and Gliessman, SR (eds.) 2016, *Agroecology in China: Science, Practice, and Sustainable Management*. CRC Press, Boca Raton, FL.

Ma, S 1987, *Agroecology Engineering in China*. Science Press, Beijing (in Chinese).

Poynter, R 2010, *The Handbook of Online and Social Media*. Wiley, Chichester, UK.

Schumilas, T 2014, "Alternative food networks with Chinese characteristics," PhD thesis, University of Waterloo.

Yang, G 2009, *The Power of the Internet in China: Citizen Activism Online.* Columbia University Press, New York.

Yang, G 2013, "Contesting food safety in the Chinese media: Between hegemony and counter-hegemony," *The China Quarterly*, 214, pp. 337–355.

2 Transformations in China's food system

*Theresa Schumilas, Zhenzhong Si,
Aijuan Chen, and Steffanie Scott*

China's food system: a hybrid of traditional and modern

Mirroring changes in the broader economy, China's food system has been described as transitional, or a hybrid of traditional and modern approaches (McCullough et al. 2008). It is a food system where a new structure of markets, different from the previous state-organized distribution, has partially evolved, but with less consolidation and integration compared to fully "modernized" food systems of the West (McCullough et al. 2008).

Changing food consumption patterns

In only three decades, China's food system has moved from one based on rationing and grain coupons to one characterized by increasing choice, rising prices (Huang, Wang, and Qiu 2012), and growing concerns about food quality and safety (Yan 2012). In 2012, China became the world's largest grocery retail market (BBC 2012). This growth is shaped by increasing urbanization and the emergence of a middle class with changing food patterns. In the 30 years of economic reform in China, between the late 1970s and 2010, per capita spending on food rose tremendously. Between 2000 and 2010 alone, total spending on food doubled, while food expenditure as a percentage of all expenditures fell from 49 per cent to 41 per cent in urban areas and from 39 per cent to 36 per cent in rural areas (Cao et al. 2013). Recent research suggests that 53 per cent of Chinese population could be categorized into lower-middle class and 7 per cent of them are upper-middle class based on annual disposable income per capita in 2015 (Leung 2016).

China's food system transition is being accompanied by a nutrition transition. Economic reforms dramatically changed food consumption patterns. Analysis of the China Health and Nutrition Survey (CHNS) reveals striking trends (Popkin 2013; Zhai et al. 2014):

- increased consumption of oils and increased frying of food;
- increased consumption of animal-sourced foods: pork remains the

most common animal-sourced food, but the intake of eggs, poultry and dairy products are all rising quickly;

- increased consumption of sugar sweetened beverages, which were non-existent prior to 1989 but have recently entered the Chinese diet as global beverage companies have expanded markets;
- decreased consumption of grains and legumes;
- increased consumption of food away from home;
- increased consumption of non-traditional foods, in particular confectionary and frozen foods.

Indeed, the only truly healthy trend revealed by the CHNS is the reduction in sodium intake, resulting from better refrigeration and therefore declining salted fish consumption. As would be expected given these trends, chronic disease rates such as obesity, diabetes, stroke, and heart disease are rising in tandem (Zhai et al. 2014; Chen et al. 2017).

Despite these negative health trends, food consumers are generally excited about more diversified food choices, greater food availability, and moving away from season-bounded choices regulated by the state (Veek, Yu, and Burns 2010). There is little evidence of anti-globalization food boycotts. Boycotts of global products or retailers that have occurred[1] have typically had a nationalistic bent, linked to the ways in which the companies have portrayed Chinese traditions in their advertising, rather than the social, ethical, or ecological concerns that characterize boycotts in the West (Dong and Tian 2009; Nyiri 2009). "Ethical" consumption in China has only recently developed. One example of this trend is vegetarian food, which is growing fast (Liu et al. 2015). Apart from this, since the late 2000s food safety has become a huge public concern and consumers are seeking better food quality, as we discuss later.

"McDonaldization," "supermarketization," "Walmartizaton," and other such processes that reflect the global experience of bigger, faster, cheaper, homogenized food products, co-exist with China's traditional food system. Yet, there is a distinct "glocalization" to their presence in China, as these global giants incorporate elements of local culture into their practices (Matusitz and Leanza 2009). The ways in which Walmart has needed to adapt to Chinese food preferences illustrates the draw of this market. For example, Walmart has catered to preferences for daily shopping and fresh foods by adding extra floor space for perishables, and tanks for customers to fish for themselves for everything from frogs and snakes to puffer fish (Matusitz and Leanza 2009). Since consumer preferences are quite regional in China, Walmart does not use central purchasing as the fresh foods in each store reflect local cuisine, necessitating sourcing from 20,000 local farms and processing firms (Matusitz and Leanza 2009). Further, China is the only country in the world where Walmart has been compelled to have a labour union (Chan 2011). After years of rapid expansion in the Chinese market, Walmart started to bet on e-commerce by collaborating with and investing in Chinese e-commerce companies in 2016. This is probably due to the increasing popularity of shopping online and the declining sales of its offline stores in recent years (Horwitz 2017). Overall, it seems that globalization of China's food system has mixed effects given the strong control maintained by the state, as well as strong culturally driven food preferences and practices. Looking at impacts of the reform and "opening" across the food chain can help unpack further details.

Food chain transition

Despite making a declining contribution to the country's overall economy, the agricultural output from China's farmers grew 4.5 times over the reform period (Huang 2011). Today, 200 million small-scale farms sell products through a complex system of formal and informal mechanisms, such as "dragon-head" enterprises, specialty cooperatives, government-run wholesalers, and uncharted systems of petty-traders to bring products from villages to diverse markets and retail formats (Garnett and Wilkes 2014; Huang 2011; Zhang and Pan 2013). Dragon-head enterprises refer to subsidized firms that hold private contracts with smallholders for specific crops and usually provide agricultural inputs to them (Huang 2011).

While the food system remains largely traditional, modernized structures and institutions are rapidly evolving. A comparison of the horticultural and livestock sub-sectors, for example, demonstrates the hybrid or transitional nature of China's food system and the co-existence of traditional and modern structures. The livestock and dairy sectors are achieving greater outputs through farm and processing consolidation. These sectors are increasingly relying on imports of feed, primarily soybeans, in order to meet growing domestic demands for animal-based foods (Huang et al. 2012; Sharma and Zhang 2014; Oliveira and Schneider 2016). The horticultural sector has also increased its productivity, so much so that, beyond meeting rising domestic demand for fruits and vegetables, China's small vegetable and fruit farmers produce for export. Output of fruits in particular grew almost 30-fold since 1980, relying not on firm and farm consolidation but rather on the complex traditional intermediaries described above (Huang et al. 2012).

There are also changes beyond the farm gate in processing. A strong domestic processing sector exists, with dairy, bakery, and dried processed foods as its leading industries (GAIN 2015). Yet, as with the production sector, the processing sector illustrates the contradictions of modern and traditional systems in co-existence. On the one hand, much of agricultural processing remains in the hands of small firms. In 2007, for example, China's food processing sector included more than 448,000 firms, of which almost 353,000 had fewer than ten employees (FORHEAD 2014). On the other hand, vertical integration, consolidation, and agricultural "modernization" is accelerating rapidly, with substantial state support. For example, large-scale enterprises with output values of over US$1.65 billion receive a special category of support, and policies have set out minimum sizes for livestock processing facilities such as abattoirs (MIIT and NDRC 2012).

Foreign direct investment (FDI) in China has catalyzed China's food processing sector. Inward FDI in the agriculture and food and beverages sectors in China more than tripled between 2004 and 2010 (Hua 2012). While this is still small in relation to the overall value of the processing sector in China, it illustrates that overseas firms are starting to play key roles in agricultural processing (e.g. ADM, Cargill, Bunge, and Wilmar), food manufacturing (e.g. Nestle, General Mills, Coca-Cola, Pepsico, Danone, Heineken), and food services (e.g. Yum! Foods, McDonald's) (Garnett and Wilkes 2014).

Food retail transitions

The food retail sector in China also illustrates the co-existence of traditional and modern structures. While wet markets and traditional marketing chains remain dominant, in the 1990s and 2000s domestic and foreign supermarket chains in China expanded faster than elsewhere, with estimates ranging between 10 per cent and 30 per cent growth per year depending on the region (Reardon et al. 2003; Hu et al. 2004). In China's largest cities, supermarkets provide most of the processed foods, rice, and dairy foods, but less of meat, fruits, and vegetables (Garnett and Wilkes 2014). Therefore, while supermarkets are expanding, most consumers still prefer to purchase foods, especially vegetables and sometimes meats and fish, in traditional wet markets, where petty-traders resell food purchased from large wholesale markets (Cui 2011; Maruyama and Wu 2014; Si et al. 2018). Nonetheless, food chains for many products are elongating (Ho 2005), largely because of state intervention. In the 2000s the state began to replace state-run wet markets with privatized wet markets (Zhang and Pan 2013) and modern supermarkets in efforts to improve hygiene by ensuring public health and labelling standards, adding toilets and washrooms, and upgrading storage and display facilities (Zhang and Pan 2013). It is interesting to note that modern retail approaches frequently embrace traditional styles of vending. Supermarkets and hypermarkets,[2] for example, have large produce and seafood sections where consumers can closely inspect food as in a market, cater to consumer demand for local cuisine and specialties by using local suppliers, and sometimes include market stalls into the design of the store (Garnett and Wilkes 2014). Out-of-home consumption and food delivery from restaurants is expanding fast, and helps to explain the slowing down of growth rates in grocery store purchases since 2014 (Lannes et al. 2017). Convenience purchases (through mini-marts) and online purchases are growing fast, while the growth of hypermarkets fell by 2 per cent between 2014 and 2016. E-commerce (especially common for snack food and other packaged foods) grew by more than 52 per cent in value over this two-year period (Lannes et al. 2017).

Agrarian transition and agroindustrialization in China

In past centuries, China's food production system was largely subsistence-based. Farmers did not organize themselves primarily for economic purposes, although farmers' organizations emerged for security and self-defence purposes during particular time periods (e.g. the war of resistance against Japan in the 1930s) (Thaxton 1997). In the socialist period, from the 1950s to the late 1970s, the collectivization campaign completely restructured the peasantry through the establishment of people's communes (*renmin gongshe*). In the process of collectivization, means of production including land were transferred from rural households to the commune, food production activities were collectively organized by brigades (*shengchan dadui*, a sub-unit of the commune), and food and income from operating collective enterprises deducted by taxes and brigade public spending was allocated according to work contributions of each household (Nolan 1983).

Although collective brigades replaced households as units of agricultural production, agricultural production remained non-commoditized in the sense that it focused on self-provisioning and direct non-monetary exchanges between units of production and with state agencies (Friedmann 1980). Since the 1950s, the state has procured food from farmers at lower than market prices and sold industrial products at a relatively higher price. The revenue from this unequal exchange of industrial and agricultural products, often called a "scissors effect," was directed to support industrial development. In this way, the agricultural sector contributed roughly 17 per cent of the total national income to the development of industries from 1950s to 1970s (Ying 2014: 86).

The collectivization of agriculture not only supplied primitive accumulation for the early industrialization of the People's Republic of China but also led to the crisis of the Chinese peasantry. The severe shortage of food and the stagnation of China's agricultural output in the 1970s led to a fundamental reform of the farmland proprietary system, known as the Household Responsibility System (HRS). The HRS propelled a process of de-collectivization across rural China through allocating rural land use rights (other than on state-owned farms) directly to rural households, according to the number of household members. Under the HRS, rural households again became the basic unit of agricultural production; rural land is mainly owned and controlled de facto by "rural collectives" that allocate and reallocate land to each household under land leasing contracts, typically for 30 years,[3] and rural land cannot be sold or purchased. Individual households have the right to use, sub-lease, and transfer land, but they do not have the land ownership. This rural institutional reform led to a rapid growth in agricultural productivity in the 1980s (Lin 1992). It also set the stage for the direct or indirect involvement of small-scale farmers in China's organic agriculture sector.

In the early 1990s, the Chinese government started to formulate and implement an agricultural modernization program with the goal of transforming China's small-scale, household-based agricultural production into modernized agricultural production, with an emphasis on market-oriented, large-scale, specialized production of higher-value goods (Zhang 2012). The Chinese government has promoted agricultural integration mainly through two parallel processes: scaling up agricultural production and integrating all activities within the supply chain (e.g. production, processing, and marketing) (Zhang and Donaldson 2008). The commercialization of agriculture contributes to two recent developments in China's agricultural sector. First, scaled-up, specialized, commercialized and vertically integrated agricultural production has in recent years been actively promoted by the Chinese government through encouraging private and state-owned companies to enter the agricultural sector (Waldron et al. 2006). Second, in this reorganization of agricultural assets (labour, land, and capital), capital, and labour have become increasingly mobile (Brauw et al. 2002; Zhang and Donaldson 2010).

The commercialization of agriculture has created increasing demand for improved technical services, specialized knowledge, and better access to marketing and technical information (World Bank 2006). Within studies of agrarian transition, many scholars have investigated challenges for small-scale farmers in dealing with global agricultural markets. These challenges include high unit transaction costs, low market competitive capacity, stringent requirements on food safety and quality, lack of bargaining power, lack of access to technologies and marketing information, etc. (Kherallah et al. 2002; Thorp et al. 2005; Poulton et al. 2005; Gulati et al. 2007; Devaux et al. 2009; Markelova et al. 2009). Poulton et al. (2005) highlight the challenges faced by small-scale farmers in quality-conscious and niche markets such as organic or fair trade. Barrett et al. (2001) argue that the high cost of third-party certification can be or is often a major barrier for small-scale farmers to participate in these markets. In China, small-scale agricultural production has also posed a great challenge for the government in monitoring production practices for food safety in a sector composed of 200 million farming households (Zhou and Jin 2009).

The process of China's agrarian transition has occurred mainly through two methods: "vertical integration" (*zongxiang yitihua*) and "agroindustrialization" (*chanyehua*). In practice, various farming operation structures have emerged and gradually replaced small-scale farming structures (Schneider 2015). The dragon-head enterprise model and the farmers' cooperative model are the most common organizational structures adopting integration strategies in China's agricultural sector (Zhang and Donaldson 2008; Huang 2011). Dragon-head enterprises,[4] "clustered groups to which state capital can be channelled and state preferential treatment provided" (Chan 2009: 46), have been promoted by the Chinese government since 2000 to overcome urban–rural income disparities and address the "three-dimensional rural issues" (*sannong wenti*) in China: rural development challenges affecting peasants, rural society, and agriculture.[5]

Food policy regime around national food security

This "agricultural modernization" that we explain above has been shaped by a complex food policy regime that centres on national food security—which is interpreted as grain productivity. Lester Brown's provocative 1994 book *Who Will Feed China?* caused quite a stir among Chinese policymakers. It came out at a time of rampant conversion of agricultural lands. Shortly after, the Chinese government established a renewed emphasis on agricultural land protection and food system localization. The 1998 revision of China's land management law gave much stronger emphasis to food security and preservation of farmland. Moreover, China initiated an approach to food system planning of having provincial governors coordinate the grain supply (encouraging procurement from within their own province where feasible) and city mayors coordinate vegetable supply for their local areas. As a result, analysts have turned to China as an insightful example of food self-sufficiency and city-scale food security (Lang and Miao 2013). As one example, in 2007, Nanjing—a city with of eight million people, including two million "rural" inhabitants and 2,366 square kilometres of cultivated land (i.e. 36 per cent of the city's total area)—supplied 44 per cent of its grain crops from within its administrative boundaries; 40 per cent of its vegetables; 20 per cent of its pork; 10 per cent of fisheries; 30 per cent of poultry; and 15 per cent of eggs (Lang and Miao 2013: 12).

Food security, or, more precisely in the Chinese context—grain productivity, has always been the pivotal food policy adherence in China. In fact, the whole system of Chinese food policy could be termed as a "food security-centred" policy regime. It was emphasized by the former central government led by the former premier Wen Jiabao, who received the Agricola Medal presented by FAO for his "life-long dedication to promoting food security and poverty reduction in the People's Republic of China, and in the world" (FAO 2012). It is clearly prioritized and highlighted in governmental agendas of fostering rural development and maintaining social stability and state sovereignty.

The priority of food security in China's national public policy system is closely associated with its historic and social background. The traditional belief in the dynastic cycle illustrates the philosophy that securing the food supply, protecting against famine, and maintaining harmony are essential for the state to maintain political legitimacy (Zha and Zhang 2013). Food security, in particular grain self-sufficiency, is embedded in Chinese political legitimacy. Providing sufficient food is the way in which political support is solidified. The state faced a legitimacy crisis linked to food insecurity at the start of the reform period when 250 million out of 800 million rural residents were impoverished and hunger was widespread (Zha and Zhang 2013). Famine and hunger are deeply ingrained in people's memories, perhaps more so than any other civilization (Li 2007). To illustrate, in the 1920s at least 500,000 people starved to death and almost 20 million were left destitute. In the 1940s, between two million and three million people died in famines in Henan province alone. Only a few years later, between 1959 and 1961, starvation during the "Great Leap Forward" killed an estimated 30 million more people (Zha and Zhang 2013).

This political philosophy mingles with people's memories of famines and results in a generally heightened importance of food in China (Simelton 2011, Tong 2011). As an old Chinese saying goes, "Food is God" (*min yi shi wei tian*). Indeed, the standard Chinese greeting, rather than "hello," is *Ni chi le mei you?* or literally, *Have you eaten yet?*, reflecting a history of food insecurity and the central place of food in society (Zhang et al. 2006). With a population increasingly demanding dietary diversity and more meat, coupled with pressures on land from increasing urbanization, food security is an ever-present concern in China and has continued as a state priority since ancient times (Carter, Zhong, and Zhu 2012; Li 2012; Zha and Zhang 2013). China's commitment to food security is also driven by the state's interest in protecting farmers in a sector that still employs more than one-third of the labour force and to address rural–urban inequities (Zha and Zhang 2013). Domestically produced grain can't compete with heavily subsidized grain from industrialized countries, so the state seeks to buffer its farmers from price downturns to ensure political stability (Zha and Zhang 2013).

China's perception of its image in the world also motivates its food security policies. The international socialism–capitalism confrontation after the founding of the People's Republic of China in 1949 entails the importance of maintaining state sovereignty in every economic sector. Producing enough food has become the first step stone of achieving this sovereignty. The state remembers the economic sanctions it experienced in 1959, when its people were suffering the worst famine in recorded history, and remains distrustful of reliance on an international food regime (Tong 2011). Since its accession to the WTO in 2001, there has been even stronger attention to food security by the Chinese state. China recognizes, as the world's largest food consumer, that reliance on trade would destabilize global markets considerably, and this has reinforced its food security policies (Wong and Huang 2012). Starting in the 1990s, to counter suggestions that it would destabilize global food prices if imports increased (Brown 1996), China responded with a range of policy instruments to stimulate agricultural production (Huang et al. 2011). Policy actions to ensure it met food security targets included the "governor's responsibility for grain

Table 2.1 Components of China's national food security policy system

Farmland preservation	Prime farmland preservation policy Land consolidation and reclamation "Dynamic balance (no net loss) of total cultivated land"
Food reserve and distribution	Enhancing the food market mechanism National food reserve system Policy support for establishing national food distribution network

Agricultural policy support	Governor and mayor responsibilities for food
	   production
	Investment in agricultural infrastructure
	   construction
	Grain prices protection
	Agricultural financial support
	Reduction of agricultural taxes and fees
	Food production subsidies
	Transfer payments to agriculture

Source: National Development and Reform Commission (2008).

productivity" and the "mayor's responsibility for vegetable productivity," which obliged provinces and cities to endorse food security targets and meet grain and vegetable quotas (Huang et al. 2011). While initially successful, these measures proved insufficient as yields began to drop in the late 1990s. In response, the state spent US$21 billion on a new series of economic policies including reducing agricultural taxes and subsidizing chemical inputs. Since that time, China has met its 95 per cent grain self-sufficiency[6] targets across all food categories (Carter, Zhong, and Zhu 2012; Morton 2012).

The priority of food security among China's public policies is underpinned by specific policies that include not only policies in agriculture production but also policies in land resource management and food reserves. Table 2.1 provides a list of important policies under the "food security" umbrella. These policies are also the reason why China can still maintain a high level of food self-sufficiency despite importing more than 80 per cent of its soybeans (in 2015) and a large amount of other major crops (Jamet and Chaumet 2016). Nevertheless, this has provoked a heightened anxiety and public debate about the nation's food security.

Farmland preservation policies

The first group of national food policies relate to farmland preservation. The efforts of the Chinese government to preserve cultivated land is mirrored by its stringent policy of "prime farmland preservation" in land use planning at multiple administrative levels. This policy prohibits any conversion of the 1.8 billion *mu* (120 million hectares) of prime farmland without the approval of the State Council. Despite this, the rapid loss of farmland and farmers due to industrialization and urbanization is a major challenge confronted by the Chinese government. It is also the biggest conflict within the food security policy regime: maintaining food productivity while also maintaining a high rate of economic development from industrialization and urbanization (see Lichtenberg and Ding 2008, 2009). Farmland conversion to industrial and residential use, both legal and illegal (see Wang and Scott 2008), has been taking place across the country and is treated by the government as a major threat to food security[7] (Lichtenberg and Ding 2008). Meanwhile, maintaining a high rate of economic growth with current economic development patterns clearly implies more farmland loss. Therefore, the Chinese government established another policy—"dynamic balance (no net loss) of total cultivated land," which is essentially a land reserve system that requires local government and land users to reclaim the same amount of farmland as they take up. To achieve this goal, an enormous governmental investment in rural land consolidation and reclamation was initiative. This vast ongoing governmental campaign, although not widely recognized by Western scholars, is demanding more farmland from rural areas by consolidating rural residential land, and is dramatically transforming the landscape and spaces of rural China. Another conflict comes from the land conservation projects. Despite the fact that part of the farmland loss is a result of land degradation, the Chinese leadership has curtailed land conservation projects to save more cultivated land (Xu et al. 2006).

Food reserve and distribution policies

The second group of policies concerns food reserves and distribution. Unlike many other countries, China is currently maintaining a large grain reserve (including 55 million tons of wheat, according to the FAO), as well as other food, such as pork and edible oils (Harkness 2011). Reflected in the old Chinese adage "where there is grain in hands, there is calmness in hearts," food reserves have a long tradition in Chinese history. The modern food reserve institution was originally established in 1990 to solve the farmers' difficulty of selling surplus grains. However, after nearly three decades, it has become an important institutional mechanism to cope with food price hikes and maintain domestic food security by adjusting and controlling the food market (see United Nations 2010). The national food reserve system now includes food reserves of central government, local governments, and food enterprises. The central government has the right and responsibility to manage the system.

Additional food security polices

Besides farmland preservation and food reserves, many other agriculture policies have contributed to the food security of China. These include, but are not limited to, public investment in agricultural research and farmland infrastructure construction, protection of grain prices by setting a minimum price, public procurement and the food reserve system, financial support and other services for agricultural production, exemption of agricultural taxes and fees, and various forms of food production subsidies and other transfer payments to agriculture. All the policies above have the same goal: to guarantee the nation's food productivity. As two examples of this, the "governor's responsibility for grain productivity" and the "mayor's responsibility for vegetable productivity," are codified into the system for assessing achievements of government officers. They are designated by the central government to guarantee a certain degree of regional food self-sufficiency. To achieve this goal, even municipal governments such as Shanghai that face severe urbanization pressures have designed corresponding policy tools such as agricultural land use planning to establish agricultural development and farmland preservation goals.

The legacy of this productivist "miracle"

The Chinese government's unequivocal embrace of productivism has resulted in negative environmental impacts. China's success in meeting its established food production targets has been primarily due to the extensive use of modern inputs, which many consider unsustainable. China's food security–driven productivist legacy is only just being revealed, and the very changes that helped production soar to meet food security goals may now be posing barriers to meeting those goals in the future. While data sources are few and unreliable, a robust body of research is beginning to reveal the extensive impacts associated with both industry and industrialized agriculture (McBeath and McBeath 2010; Holdaway 2013; Gilley 2012).

Use of synthetic fertilizer has expanded fivefold since reforms began, and China now leads the world in both the production and the use of synthetic fertilizers (Fan et al. 2012). Chemical fertilizer use increased from 8.8 to 54 million metric tons between 1978 and 2009, and applications per hectare increased from 59 kg to 341 kg over the same period. A low fertilizer nutrient use efficiency has been noted, and there are high nutrient losses due to inappropriate application (Zhang and Shen 2013). Further, fertilization is considered unbalanced. There are large regional variations in amounts used and, while nitrogen and phosphorus have generally been over-used, potassium use has been insufficient in comparison and shows declining balances in multiple soils across China (Zhang and Shen 2013). This overuse of nitrogen-based fertilizer in particular has resulted in eutrophication of surface water, excessive greenhouse gas emissions (Zhang and Shen 2013), and soil acidification in multiple regions (Holdaway and Hussain 2014).

Parallel increases in other agricultural chemicals (herbicides, fungicides, pesticides), as well as agricultural plastics, have been noted since the 1970s

(Carter et al. 2012). The use of pesticides increased 2.4 times between 1990 and 2010 to over 17 million tons (Holdaway 2013; Fan et al. 2012), making China the world's second largest producer and consumer of pesticides, responsible for nearly 35 per cent of all global consumption (Zhang and Shen 2013). Excessive pesticides persist in soil and some 16 million hectares of cropland in China are estimated to be polluted by agricultural pesticides (Tóth and Li 2013).

Beyond land pollution, loss of organic matter and soil erosion are critical problems. Across the country, 38 per cent of the soil suffers from nutrient and organic matter losses associated with erosion (Fan et al. 2012). For example, the average soil organic matter in topsoil in China is 10 g/kg, compared to 25–40 g/kg in Europe and the US (Holdaway and Husain 2014). Further, heavy metal contamination of soil, as a result of rural industrial processes, pesticides, and manure, is suspected to be a problem that is more severe than in many other countries, although data are scarce and suspect (Holdaway and Husain 2014).

Food safety crisis and loss of trust

The growing public anxiety around food safety in recent years constituted a critical part of the broad transformation of China's food system and played a pivotal role in the development of the ecological agriculture sector. Chinese consumers understand food safety broadly to include not only food produced under sanitary conditions, but also food that is unadulterated by illicit food additives and free from environmental pollutants and agricultural inputs such as antibiotics and pesticides (Holdaway and Husain 2014; Yang 2013). China's food safety scandals started to receive exponential attention, by scholars and the global press, in 2008, when 40,000 infants had to be hospitalized because of deliberate contamination of milk powder with melamine (Yan 2012). Since that time, scholars have begun to unpack the ways China's continuing food safety scandals reveal deep social and political processes that constitute a "food safety crisis" (Cheng 2012; Yan 2012). Yan (2012) proposed a typology to help understand the various incidents that comprise the "crisis." He notes that incidents can be divided into three semi-distinct types. First, food hygiene problems, common in pre-modern China, have continued despite a more industrialized food system. Second, Yan (2012: 707) describes a category of "unsafe foods," which are generally incidents associated with the extensive use of fertilizers and pesticides in China's industrialized food sector. Many incidents fall into this category. Reports suggest that nearly 50 per cent of fruits and vegetables in China have pesticide residues exceeding official standards and that each year more than 100,000 people become sick due to pesticide exposure (Holdaway 2013; Yang 2007). Third, Yan (2012: 710) characterizes some food safety problems as "poisonous foods" and the types of scandals that have most provoked the food safety crisis.

Poisonous foods are a newer phenomenon in China, and can be differentiated from other types of food safety problems because they involve deliberate contamination and thus serious ethical concerns (Yan 2012). There are multiple pathways through which food is intentionally contaminated by processors and producers to boost profits. Harmful inputs include the use of dye to colour vegetables and berries to improve their appearance. Dye has also been fed to poultry so egg yolks are more brightly coloured. Cooking oil has been reclaimed and adulterated. And, in one of the highest-profile cases, melamine, a nitrogen-rich organic compound, has been added to milk to boost its protein content cheaply (Yan 2012). Yan (2012) also describes a category of "fake foods," which present "a challenge to the imagination" (p. 712). Examples are staggering and nauseating, and include starch masquerading as milk powder, soy sauce made from human hair, and chicken eggs made from water and assorted chemicals. These cases of deliberate food adulteration in the pursuit of profit are most disturbing because evidence suggests they all too frequently occur with the knowledge of government officials (Yan 2012; FORHEAD 2014). Hence, while research suggests the incidence of ill health from the deliberate adulteration of food may indeed be lower than from food hygiene problems, these "fake food" examples are "socially lethal" (Yan 2012: 717) because of the intention to do harm with the apparent knowledge of the state, and because of the widespread fear, panic, and distrust they engender.

A generalized distrust around food exists in Chinese society, and state efforts thus far have failed to rebuild trust in food. Our research confirms this. Every individual we spoke with initiated a conversation about China's food safety crisis, even though we never prompted this with a question. Despite harsh penalties for those convicted, and new food safety legislation and enforcement systems, the problem continues because of bureaucratic fragmentation, competition among regulatory agencies, and corruption of officials (Jia and Jukes 2013). Canadian sociologist Hongming Cheng (2012), who has investigated white-collar crime in China, turned his attention to "food crime" (p. 254). He argues that food scandals are perpetuated by the existence of a "helix of industry-government-university relations" (Cheng 2012: 257). The government's own public surveys consistently show food safety as a top concern, revealing that by the end of 2010, 18 months after the state passed more stringent food safety legislation, 70 per cent of surveyed consumers still ranked food safety as a top concern (Yan 2012). In an unprecedented move, the party-state acknowledged its inability to provide safe food to its people. A 2008 Ministry of Commerce report that admits "the increase in public concern about food safety may be an indicator of the decline of consumer confidence in the government's ability to regulate food safety" (cited in Yan 2012: 724). Indeed, state officials themselves do not trust the food supply. The previous Mao era practice of a "special supply" of food designated for government officials and intellectuals that existed when food shortages were a common occurrence, now functions to ensure safe food for these elite groups (Yan 2012: 723). There seems to be no resolution to this problem in sight.

Consumers' food safety concerns appear to be conflated with broader quality concerns and fears associated with the modern food system and industrialized production methods described as "unnatural" or "polluted" (Klein 2013: 384). Trust in food is entangled with multiple issues, such as people's understanding of place and seasonal cycles (as in debates over food grown in greenhouses), regional cuisine, and perceptions of the food vendor or provider (Klein 2013). In the face of ineffective food governance, China's food safety crisis has perpetuated a broader crisis of distrust of the market, of individuals, and of the state. This "food safety crisis" thus exists in a context in which people have been untied from collective institutions that protected them (however imperfectly) during the reform period, but they still look to the state—though perhaps more tentatively— for ensuring food security and safety. Yet, our research suggests that the same processes that are contributing to the food crisis might be stimulating diverse solutions by encouraging new connections to rebuild trust, not in the state's ability to provide safe food but rather in place-based, face-to-face relationships.

Strong state versus weak civil society

Whereas populist and democratic concerns about food safety or environmental impacts of industrial agriculture can be readily expressed in the North America and Western Europe, parallel civil society mechanisms are absent in China, where a state-controlled, elitist decision-making approach excludes people from policy advocacy processes (Gilley 2012). The approach eschews participation of either independent academia or civil society, and authoritatively proclaims scientific knowledge based on participation by technocratic elites who decide which "science" is listened to, and ignoring scholars with non-dominant views (Gilley 2012).

The state's ever-changing relationship with NGOs illustrates the sociopolitical context in which ecological agriculture is evolving. Since the 1980s, the liberal democratic notion of the separation of state and individual has grown in Chinese society and a non-government sector has exploded to tens of thousands of NGOs (Hsu 2011) in fields as diverse as education, environmental health, housing, and poverty alleviation (Spires 2011). However, scholars hasten to add that the interpretation of this expansion needs to be understood beyond mere numbers, and that the Chinese understanding of NGO and civil society is rather different from that in liberal democracies in the West, so there are many questions about what exactly "counts" as an NGO. NGOs in China are characterized by alliances with government rather than being independent institutions (Hsu 2011). Indeed, registering an NGO in China requires a government department to endorse and sponsor the initiative, making official status impossible for NGOs that seek to resist state directives.[8]

In its overarching goal of maintaining harmony, the Chinese state routinely places restrictions on NGO actions. In the face of rhetoric about "small state, big society," NGOs in China work around a "hazy, shifting boundary" (Stern and O'Brien 2012: 3) where mixed signals about what is permitted are common (2012: 3). Heilmann and Perry (2011: 12) refer to this process as "guerrilla policy making" characterized by "continual improvisation and adjustment" that creates a climate of "pervasive uncertainty" for those challenging the state. Indeed, advocacy on sensitive issues or use of particular tactics are more likely than others to land an NGO in trouble. For example, the state opposes actions that focus on demands for rights, or resistance that seems to be building cross-class or cross-locality alliances (Bruun 2013; Stern and O'Brien 2012; O'Brien and Li 2006). A common state response to groups that it considers to be destabilizing forces is to entangle them in endless bureaucracy and paperwork such as disputes about taxes or missing permits (Cai 2008).

In this chapter, we have outlined the complicated context in which developments in the organic and ecological agriculture sector are taking place. We have described a food system in transition where a new structure of markets, different from the previous state-organized distribution, has partially evolved. While food production is less consolidated than what we find in North America (indeed, 200 million small-scale farms still supply the country through a complex system of modern and traditional markets), modernized structures and institutions are rapidly evolving and "supermarketization" of food retail is happening fast (Si et al. 2018). This food system transition is being accompanied by a nutrition transition, with a "Westernizing" of diets and food consumption patterns.

The food and agriculture policies driving these transitions are equally complicated. Driven by memories of famines and a heightened cultural importance of food (compared to North America), a preoccupation with food security has reinforced a productivist approach to agriculture that is driving traditional agriculture to the margins. The ecological disaster of China's "food security miracle" is being revealed and is shaping growing public anxiety about food safety and widespread distrust that the state has been as yet unable to dispel. In the West, civil society responses to such concerns led to the development of the organic sector and safe food policies. However, as described in Chapter 3, the ever-changing restrictions placed on NGOs, together with the state-dominated policy-setting processes in China, have resulted in a state-driven (or top-down) approach to ecological and organic sector development, as we explain in the following chapter.

Notes

1 Nationalistic boycotts have been levelled against Carrefour, Coca-Cola, McDonalds, and Starbucks, for example. See Nyiri (2009) for a full discussion.
2 The term "hypermarket" typically refers to a large store that combines a food supermarket with a more general department store.
3 "Land in the countryside and in suburban areas is under collective ownership unless the law stipulates that the land is state-owned" (National People's Congress 1982, Article 10). In 2008, the 30-year contract period was extended to become a stable right that "would not change in the long term" (*changjiu bubian*) (Central Committee of the Communist Party of China 2008).
4 Agricultural dragon-head enterprises must benefit a large number of farmers and contribute to the agroindustrialization and agricultural integration process. They are classified by the Ministry of Agriculture into three levels: national, provincial, and city levels. City-level dragon-head enterprises have less stringent criteria than provincial- or national-level equivalents. The criteria relate to turnover, profits, market share, taxes paid, growth rates, and linking with a certain minimum number of small-scale farmers (MOA 2002).
5 See more discussion about the *san nong wenti* in Day (2008, 2013).
6 There is growing debate as to the degree of food self-sufficiency in China. Other scholars suggest that recently this number has fallen to 85 per cent food self sufficient. Estimates vary depending on the definition of "food security" used. In particular there is debate about whether animal feed, such as soybeans, should be considered a "grain" and thus included in the definition of "food security."
7 Lichtenberg and Ding's analysis reveal that current farmland protection policies are not the most efficient measures to maintain China's national food security.
8 A change of regulations regarding NGO registration in 2011 no longer required several types of NGOs to find a governmental supervisory entity, and enabled them to directly register with the Civil Affairs Department. Whether this change means greater or more government control of NGOs is debatable. On January 1, 2017, the Chinese government enacted a law on "the Administration of Activities of Activities of Overseas Non-Governmental Organizations within the Territory of China," which placed more restrictions on the activities of foreign NGOs.

References

Barret, HR, Browne, AW, Harris, PJC, and Cadoret, K 2001, "Smallholder farmers and organic certification: Accessing the EU market from the developing world," *Biological Agriculture and Horticulture*, vol. 19, no. 2, pp. 183–199.

BBC News 2012, "China surpasses US as world's biggest grocery market," April 4. Accessed at www.bbc.com/news/business-17595963.

Brauw, A, Huang, J, Rozelle, S, Zhang, L, and Zhang, Y 2002, "The evolution of China's rural labour markets during the reforms," *Journal of Comparative Economics*, vol. 30, no. 2, pp. 329–353.

Brown, L 1996, "Who will feed China?" *The Futurist*, vol. 30, no. 1, January, pp. 14–18.

Bruun, O 2013, "Social movements, competing rationalities and trigger events: The complexity of Chinese popular mobilizations," *Anthropological Theory*, vol. 13, no. 2, pp. 249–266.

Cai, Y 2008, "Local governments and the suppression of popular resistance in China," *China Quarterly*, no. 193, pp. 24–42.

Cao, L, Tian, W, Wang, J, Malcolm, B, Liu, H, and Zhou, Z 2013, "Recent food consumption trends in China and trade implications to 2020," *Australian Agribusiness Review*, vol. 21, pp. 14–44.

Carter, C, Zhong, F, and Zhu, J 2012, "Advances in Chinese agriculture and its global implications," *Applied Economic Perspectives and Policy*, vol. 34, no. 1, pp. 1–36.

Chan, A 2011, "Unionizing Chinese Walmart stores," In: Chan, A (ed.), *Walmart in China*. Cornell University Press, Ithaca, NY, pp. 199–217.

Chan, H 2009, "Politics over markets; integration sate-owned enterprises into Chinese socialist market," *Public Administration and Development*, vol. 29, no. 1, pp. 43–54.

Chen, W, Gao, R, Liu, L, Zhu, M, Wang, W, Wang, Y, Wu, Z, Li, H, Gu, D, Yang, Y, Zheng, Z, Jiang, L, and Hu, S 2017, "China cardiovascular diseases report 2015: A summary." *Journal of Geriatric Cardiology*, vol. 14, no. 1, pp. 1–10.

Cheng, H 2012, "Cheap capitalism: A sociological study of food crime in China," *British Journal of Criminology*, vol. 52, no. 2, pp. 254–273.

Central Committee of the Communist Party of China 2008, *Decision of the Central Committee of the Communist Party of China on Several Major Issues Concerning the Reform and Development of Rural Areas*. Accessed at http://cpc.people.com.cn/GB/64093/64094/8194418.html on July 19, 2017.

Cui, B 2011, "The choice behavior in fresh food retail market: A case study of consumers in China," *International Journal of China Marketing*, vo.2, no. 1, pp. 68–77.

Day, A 2008, "The end of the peasant? New rural reconstruction in China," *Boundary*, vol. 35, no. 2, pp. 49–73.

Day, A 2013, *The Peasant in Postsocialist China: History, Politics, and Capitalism*, Cambridge University Press, Cambridge, UK.

Devaux, A, Horton, D, Velasco, C, Thiele, G, López, G, Bernet, T, Reinoso, I, and Ordinola, M 2009, "Collective action for market chain innovation in the Andes," *Food Policy*, vol. 34, no. 1, pp. 31–38.

Dong, L, and Tian, K 2009, "The use of Western brands in asserting Chinese national identity," *Journal of Consumer Research*, vol. 36, no. 3, pp. 504–523.

Fan, M, Shen, J, Yuan, L, Jiang, R, Chen, X, Davies, WJ and Zhang F 2012, "Improving crop productivity and resource use efficiency to ensure food security and environmental quality in China," *Journal of Experimental Botany*, vol. 63, no. 1, pp. 13–24.

FAO (Food and Agriculture Organization) 2012, *Chinese Premier Wen Jiabao Given Agricola Medal, FAO's Highest Award*. Accessed at www.fao.org/news/story/en/item/161337/icode on July 26, 2017.

FORHEAD (Forum on Health, Environment and Development) 2014, *Food Safety in China: A Mapping of Problems, Governance and Research*. Accessed at http://webarchive.ssrc.org/cehi/PDFs/Food-Safety-in-China-Web.pdf on July 25, 2017.

Friedmann, H 1980, "Household productions and the national economy: Concepts for the analysis of agrarian formations," *Journal of Peasant Studies*, vol. 7, no. 2, pp. 158–184.

GAIN (Global Agricultural Information Network) 2015, *China's Food Processing Annual Report*, Report no. 15803. United States Department of Agriculture Foreign Agricultural Service, Washington, DC.

Garnett, T and Wilkes, A 2014, *Appetite for Change: Social, Economic and Environmental Transformations in China's Food System*, Food Climate Research Network.

Gilley, B 2012, "Authoritarian environmentalism and China's response to climate change," *Environmental Politics*, vol. 21, no. 2, pp. 287–307.

Gulati, A, Minot, N, Delgado, C, and Bora, S 2007, "Growth in high-value agriculture in Asia and the emergence of vertical links with farmers." In: Swinnen, JFM (ed.), *Global Supply Chains, Standards and the Poor: How the Globalization of Food Systems and Standards Affects Rural Development and Poverty.* CABI, Wallingford, UK, pp. 91–108.

Harkness, J 2011, *Food Security and National Security: Learning from China's Approach to Managing Its Wheat Supplies.* Institute for Agriculture and Trade Policy, Minneapolis, MN. Accessed at www.iatp.org/files/2011_02_28_FoodSec-NatSec-JH.pdf on July 19, 2017.

Heilmann, S and Perry, EJ 2011, "Embracing uncertainty: Guerrilla policy style and adaptive governance in China." In: Heilmann, S and Perry, E (eds.), *Mao's Invisible Hand: The Political Foundations of Adaptive Governance in China*, Cambridge, MA, Harvard University Press, pp. 1–29.

Ho, S 2005, "Evolution versus tradition in marketing systems: The Hong Kong food-retailing experience," *Journal of Public Policy and Marketing*, vol. 24, no. 1, pp. 90–99.

Holdaway, J 2013, "Environment and health research in China: The state of the field," *The China Quarterly*, vol. 214, pp. 1–22.

Holdaway, J and Hussain, L 2014, "Food safety in China: A mapping of problems, governance and research," Forum on Health, Environment and Development (FORHEAD). Accessed at http://webarchive.ssrc.org/cehi/PDFs/Food-Safety-in-China-Web.pdf on July 17, 2018.

Horwitz, Josh. 2017, *Walmart's Future in China Increasingly Depends on a Single Chinese Company.* Accessed at https://qz.com/992623/walmarts-future-in-china-increasingly-depends-on-jd-com-alibabas-only-rival.

Hsu, C 2011, "Even further beyond civil society: The rise of internet-oriented Chinese NGO," *Journal of Civil Society*, vol. 7, no. 1, pp. 123–127.

Hu, D, Reardon, T, Rozelle, S, Timmer, P, and Wang, H 2004, "The emergence of supermarkets with Chinese characteristics: Challenges and opportunities for China's agricultural development," *Development Policy Review*, vol. 22, no. 5, pp. 557–586.

Hua S 2012, "Current status, trends and responses in the use of foreign investment in China's agriculture," *World Agriculture*, vol. 4, pp. 73–77.

Huang, J, Wang, X, and Qiu, H 2012, *Small-Scale Farmers in China in the Face of Modernisation and Globalisation.* International Institute for Environment and Development (IIED) and HIVOS, London/The Hague.

Huang, J, Wang, S, Zhi, H, Huang, Z, and Rozelle, S 2011, "Subsidies and distortions in China's agriculture: Evidence from producer-level data," *Australian Journal of Agricultural and Resource Economics*, vol. 55, no. 1, pp. 53–71.

Huang, P 2011, "China's new-age small farms and their vertical integration: Agribusiness or co-ops?" *Modern China*, vol. 37, no. 2, pp. 107–134.

Jamet, J and Chaumet, J 2016, "Soybean in China: Adapting to the liberalization," *Oilseeds and Fats Crops and Lipids*, vol. 23, no. 6, pp. 1–9.

Jia, C and Jukes, D 2013, "The national food safety control system of China—A systematic review," *Food Control*, vol. 32, no. 1, pp. 236–245.

Kherallah, M, Delgado, C, Gabre-Madhin, E, Minot, N, and Johnson, M 2002, *Reforming Agricultural Markets in Africa*. Johns Hopkins University, Baltimore, MD.

Klein, J 2013, "Everyday approaches to food safety in Kunming," *The China Quarterly*, vol. 214, pp. 376–393.

Lang, G and Miao, B 2013, "Food security for China's cities," *International Planning Studies*, vol. 18, no. 1, pp. 5–20.

Lannes, B, Ding, J, Kou, M, Yu, J 2017, "China's two-speed growth: In and out of the home," *Insights China Shopper Report* (Bain Company), vol. 1. Accessed at www.bain.com/publications/articles/china-shopper-report-2017-chinas-two-speed-growth-in-and-out-of-the-home.aspx.

Leung, A 2016, *The Chinese Consumer in 2030*. Economist Intelligence Unit. Accessed at www.eiu.com/public/topical_report.aspx?campaignid=Chineseconsumer2030 on April 13, 2018.

Li, J 2012, "Fight silently: Everyday resistance in surviving state owned enterprises in contemporary China," *Global Labour Journal*, vol. 3, no. 2, pp. 194–216.

Li, L 2007, *Fighting famine in North China*, Stanford University Press, Stanford, CA.

Lichtenberg, E and Ding, C 2008, "Assessing farmland protection policy in China," *Land Use Policy*, vol. 25, no. 1, pp. 59–68.

Lichtenberg, E and Ding, C 2009, "Local officials as land developers: Urban spatial expansion in China," *Journal of Urban Economics*, vol. 66, no. 1, pp. 57–64.

Lin, JY 1992, "Rural reforms and agricultural growth in China," *The American Economic Review*, vol. 82, no. 1, pp. 34–51.

Liu, C, Cai, X, and Zhu, H 2015, "Eating out ethically: An analysis of the influence of ethical food consumption in a vegetarian restaurant in Guangzhou, China," *Geographical Review*, vol. 105, no. 4, pp. 551–565.

Markelova, H, Meinzen-Dick, R, Hellin, J, and Dohrn, S 2009, "Collective action for smallholder market access," *Food Policy*, vol. 34, no. 1, pp. 1–7.

Maruyama, M and Wu, L 2014, "Quantifying barriers impeding the diffusion of supermarkets in China: The role of shopping habits," *Journal of Retailing and Consumer Services*, vol. 21, no. 3, pp. 383–393.

Matusitz, J, and Leanza, K 2009, "Wal-Mart: An analysis of the glocalization of the Cathedral of Consumption in China," *Globalizations*, vol. 6, no. 2, pp. 187–205.

McBeath, J and McBeath, G 2010, *Environmental Change and Food Security in China*. Springer, Dordrecht.

McCullough, E, Pingali, P, and Stamoulis, K 2008, "Small farms and the transformation of food systems: an overview," In: McCullough, E, Pingali, P, and Stamoulis, K (eds.), *The Transformation of Agri-Food Systems: Globalization, Supply Chains and Smallholder Farmers*. Earthscan, London, pp. 3–46.

Ministry of Industry and Information Technology (MIIT) and the National Development and Reform Commission (NDRC) 2012, *Twelfth Five-Year Plan for the Food Processing Industry*. Beijing.

MOA (Ministry of Agriculture, PRC) 2002, "The designation of agricultural industrialization dragon-head enterprises." Accessed at www.moa.gov.cn/zwllm/gzgf/200307/t20030702_96674.htm on October 18, 2014.

Morton, K 2012, *Learning by Doing: China's Role in the Global Governance of Food Security*. Working Paper, Indiana University, Indiana.

National Development and Reform Commission (NDRC) 2008, "National Food Security Mid- and Long-term Planning Outline." Accessed at www.gov.cn/jrzg/2008-11/13/content_1148372.htm (in Chinese).

National People's Congress 1982, "Constitution of the People's Republic of China (1982)." Accessed at www.npc.gov.cn/wxzl/wxzl/2000–12/06/content_4421.htm on July 19, 2017.

Nolan, P 1983, "De-collectivisation of agriculture in China, 1979–82: A long-term perspective," *Economic and Political Weekly*, vol. 18, no. 32, pp. 1395–1406.

Nyiri, P 2009, "From Starbucks to Carrefour: Consumer boycotts, nationalism and taste in contemporary China," *PORTAL Journal of Multidisciplinary International Studies*, vol. 6, no. 2, pp. 1–25.

O'Brien, K and Li, L 2006, *Rightful Resistance in Rural China*, Cambridge University Press, New York.

Oliveira, G and Schneider, M 2016, "The politics of flexing soybeans: China, Brazil and global agroindustrial restructuring," *The Journal of Peasant Studies*, vol. 43, no. 1, pp. 167–194.

Organisation of Economic Co-operation and Development and Food and Agriculture Organization (OECD-FAO) 2013, *Agricultural Outlook 2013–2022*. OECD-FAO, Rome.

Popkin, B 2013, "Synthesis and implications: China's nutrition transition in the context of changes across other low- and middle-income countries," *Obesity Reviews*, vol. 15, no. 1, pp. 60–67.

Poulton, C, Dorward, A, and Kydd, J 2005, *The Future of Small Farms: New Directions for Services, Institutions and Intermediation*, paper presented at the Future of Small Farms Workshop, Imperial College, Wye, UK (June 26–29).

Reardon, T, Timmer, C, Barrett, C, and Berdegue, J 2003, "The rise of supermarkets in Africa, Asia and Latin America," *American Journal of Agricultural Economics*, vol. 85, no. 5, pp. 1140–1146.

Schneider, M 2015, "What, then, is a Chinese peasant? Nongmin discourse and agroindustrialization in contemporary China," *Agriculture and Human Values*, vol. 32, no. 2, pp. 331–346.

Scott, S, Si, Z, Schumilas, T, and Chen, A 2014, "Contradictions in state- and civil society-driven developments in China's ecological agriculture sector," *Food Policy*, vol. 45, pp. 158–166.

Sharma, S and Zhang R 2014, *China's Dairy Dilemma: The Evolution and Future Trends of China's Dairy Industry*. Institute for Agriculture and Trade Policy, Minneapolis, MN.

Si, Z, Scott, S, and McCordic, C 2018, "Wet markets, supermarkets and alternative food sources: Consumers' food access in Nanjing, China," *Canadian Journal of Development Studies*.

Simelton, E 2011, "Food self-sufficiency and natural hazards in China," *Food Security*, vol. 3, no. 1, pp. 35–52.

Spires, A 2011, "Contingent symbiosis and civil society in an authoritarian state: Understanding the survival of China's grassroots NGOs," *American Journal of Sociology*, vol. 117, no. 1, pp. 1–45.

Stern, R and O'Brien, K 2012, "Politics at the boundary: Mixed signals and the Chinese state," *Modern China*, vol. 38, no. 2, pp. 1–25.

Thaxton, R 1997, *Salt of the Earth: The Political Origins of Peasant Protest and Communist Revolution in China*, University of California Press, Berkeley, CA.

Thorp, R, Stewart, F and Heyer, A 2005, "When and how far is group formation a route out of chronic poverty?" *World Development*, vol. 33, no. 6, pp. 907–920.

Tong, Y 2011, "Morality, benevolence and responsibility: Regime legitimacy in China from past to the present," *Journal of Chinese Political Science*, vol. 16, no. 2, pp. 141–159.

Tóth, G and Li, X (eds.), "Threats to the soil resource base of food security in China and Europe," Sino-EU Panel on Land and Soil, Accessed at http://citese-erx.ist.psu.edu/viewdoc/download?doi=10.1.1.397.992&rep=rep1&type=pdf on July 17, 2018.

United Nations 2010, "Mandate of the special rapporteur on the right to food." Accessed at www2.ohchr.org/english/issues/food/docs/CHINA_food_prelimi-nary_conclusions.doc on September 10, 2017.

Veek, A, Yu, H, and Burns, A 2010, "Consumer risks and new food systems in urban China," *Journal of Macromarketing*, vol. 30, no. 3, pp. 222–237.

Waldron, S, Brown, C, and Longworth, J 2006, "State sector reform and agriculture in China," *China Quarterly*, no. 186, pp. 277–294.

Wang, Y and Scott, S 2008, "Illegal farmland conversion in China's urban periphery: Local regimes and national transitions," *Urban Geography*, vol. 29, no. 4, pp. 327–347.

Wong, J and Huang, Y 2012, "China's food security and its global implications," *China: An International Journal*, vol. 10, no. 1, pp. 13–124.

World Bank 2006, *Enhancing Agricultural Innovation: How to Go Beyond the Strengthening of Research System*, The World Bank, Washington, DC.

Xu, Z, Xu, J, Deng, X, and Huang, J 2006, "Grain for Green versus grain: Conflict between food security and conservation set-aside in China," *World Development*, vol. 34, no. 1, pp. 130–148.

Yan, Y 2012, "Food safety and social risk in contemporary China," *The Journal of Asian Studies*, vol. 71, no. 3, pp. 705–729.

Yang, G 2013, "Contesting food safety in the Chinese media: Between hegemony and counter-hegemony," *The China Quarterly*, vol. 214, June 2013, pp. 337–355.

Yang, Y 2007, *A China Environmental Health Project Factsheet: Pesticides and Environmental Health Trends in China.* Accessed at www.wilsoncenter.org/topics/docs/pesticides_feb28.pdf on June 20, 2011.

Ying, X 2014, *Chinese Society.* China Renmin University Press, Beijing (in Chinese).

Zha, D and Zhang, H 2013, "Food in China's international relations," *The Pacific Review*, vol. 26, no. 5, pp. 455–479.

Zhai, F, Du, S, Wang, Z, Zhang, J, Du, W, and Popkin, B 2014, "Dynamics of the Chinese diet and the role of urbanicity, 1991–2011," *Obesity Reviews*, vol. 15, no. 1, pp. 16–26.

Zhang, D, Jim, C, Lin, G, He, S, Wang, J, and Lee, H 2006, "Climatic change, war and dynastic cycles in China over the last millennium," *Climatic Change*, vol. 76, no. 3–4, pp. 459–477.

Zhang, G and Shen, R 2013, "Impact of high intensity land uses on soil and environment in China," In: Toth, G and Xiubin, L (eds.), *Threats to the Soil Resource Base of Food Security in China and Europe. A Report from the Sino-EU Panel on Land and Soil.* Luxemburg: Office of the European Union, pp. 53–90.

Zhang, Q and Donaldson, J 2008, "The rise of agrarian capitalism with Chinese characteristics: Agricultural modernization, agribusiness and collective land rights," *The China Journal*, no. 60, pp. 25–47.

Zhang, Q and Donaldson, J 2010, "From peasants to farmers: Peasant differentiation, labor regimes, and land-rights institutions in China's agrarian transition," *Politics and Society*, vol. 38, no. 4, pp. 458–489.

Zhang, Q and Pan, Z 2013, "The transformation of urban vegetable retail in China: Wet markets, supermarkets and informal markets in Shanghai," *Journal of Contemporary Asia*, vol. 43, no. 3, pp. 497–518.

Zhang, QF 2012, "The political economy of contract farming in China's agrarian transition," *Journal of Agrarian Change*, vol. 12, no. 4, pp. 460–483.

Zhou, J and Jin, S 2009, *Adoption of Food Safety and Quality Standards by China's Agricultural Cooperatives: A Way Out of Monitoring Production Practices of Numerous Small-Scale Farmers?* Contributed paper prepared for presentation at the International Association of Agricultural Economists Conference, Beijing, China, August 16–22, 2009. Accessed at http://ageconsearch.umn.edu/handle/50293 on April 6, 2014.

3 Top-down initiatives

State support for ecological and organic agriculture in China

Aijuan Chen, Steffanie Scott, and Zhenzhong Si

Evolutions in Chinese ecological agriculture (CEA) and recent growth

Around the world, growing awareness of health and environmental issues associated with the intensive use of chemical inputs has spurred interest in alternative forms of agricultural production, including organic agriculture (Browne et al. 2000). Certified organic agriculture is practised in 179 countries worldwide, on 51 million hectares of land (including conversion areas) (Willer and Lernoud 2017a, 2017b). The global market for organic food was estimated to have reached US$81.6 billion (approximately €75 billion) by 2015 (Organic Monitor 2017).

Although the organic sector as a whole is still tiny, accounting for only 1.1 per cent of global agricultural land, it has been one of the fastest-growing sectors in recent decades, with double-digit annual growth in land under organic cultivation, value of organic produce and number of organic farmers (Willer and Lernoud 2017a, 2017b). Organic agriculture has taken different development paths in different countries. Marketing channels and ownership structures of organic agriculture have been diversifying, alongside the increasing demand for organic products domestically (Egelyng et al. 2010; Menon, Sema, and Partap 2010; Osswald and Menon 2013).

The emergence of ecological agriculture in China was closely associated with the severe environmental degradation that resulted from several decades of agricultural industrialization. The problems of the industrialization of farming approaches stem from a variety of practices: the expansion of farming into ecologically vulnerable areas; the overuse of synthetic fertilizer, pesticides, herbicides, fungicides, and other chemicals (Zhu and Chen 2002; Kahrl et al. 2010; Yang 2012; Li et al. 2014; Zhang et al. 2015); the excessive use of plastic film and greenhouses; the exploitation of soil and water resources (Wang et al. 2009; Gong et al. 2013); and animal waste that has polluted the water system (Li 2013; IPES-Food 2016). The degradation of agricultural resources and the environment has led to severe food safety problems, as we discuss in other chapters.

In response to the severely degraded environment and increasing demand for safer food (detailed in Chapter 2), the Chinese government has, since the late 1980s, sought ways to promote a more environmentally sustainable agricultural system. The "Chinese Ecological Agriculture" (CEA) initiatives of the 1980s had three main goals: food security, employment and income generation, and natural resource conservation and environmental protection (Sanders 2000; Shi 2002). By 1990, approximately 1,200 "pilot ecological agriculture villages," or eco-villages, had been established (Zong 2002: 54). Major achievements of the CEA initiative include higher agricultural productivity with fewer external inputs, farming systems with higher resilience during natural disasters, and improved rural landscapes (Shi 2002). Despite the success of some pilot projects, CEA was not expanded on a larger scale across rural China. It is no longer being promoted in China owing to many challenges: limited knowledge of CEA among farmers; low incentives due to the lack of price premiums on ecological food products; and the scheme being driven and dominated by village leaders rather than farmers (see Sanders 2000). Nevertheless, CEA marked the first national attempt to address environmental issues in the countryside through promoting sustainable agricultural practices. It helped to make the concept of ecological agriculture more widely recognized in China (Zhao et al. 2008), which contributed to the new wave of agriculture greening practices that followed it its wake.

Legislating ecological and organic agriculture in China

The promotion of sustainable and ecological agriculture has been highlighted in the national sustainable development strategy and China's No. 1 Central Documents, which set out annual policy priorities (Zhao et al. 2008; Liu, Pieniak, and Verbeke 2013). In 2015, eight ministries and bureaus including the then Ministry of Agriculture[1] and Ministry of Environmental Protection issued the "National Sustainable Agriculture Development Plan (2015–2030)." It laid out the opportunities and challenges, key principles, developmental goals, and major tasks and projects, as well as institutional schemes and arrangements to achieve the sustainability of agriculture before the year 2030. The theme of environmental sustainability of agriculture has been appearing frequently in various governmental documents and announcements. The No. 1 Central Government Document in 2017 listed "promoting green agricultural production approaches, enhancing the capacity of sustainable development of agriculture" as the second major developmental goal.

The frequent emphasis on sustainable agriculture in policies in recent years is a continuation of more than two decades of efforts to institutionalize ecological and organic agriculture. To deal with severe environmental degradation and various food safety issues, a system of progressively stringent food quality production standards has been in place in China since the 1990s to meet various demands for domestic and international markets. These standards include green food (*lüse shipin*), hazard-free food (*wugonghai shipin*), and organic food (*youji shipin*). Compared with the organic production standards, green food and hazard-free food quality standards are less stringent. They were created by the Chinese government with China's conditions of a heavily polluted environment and relatively poorly educated farmers in mind (Paull 2008). By introducing ecolabelling, the Chinese government sought to increase consumer confidence in food safety, reduce negative environmental impacts, and improve farmers' incomes (Giovannucci 2005). These initiatives have shaped China's path toward the greening of agriculture.

Green food

In 1990, the Ministry of Agriculture created a green food program in response to increasing concerns regarding both environmental degradation and unsafe food (Sanders 2006). The China Green Food Development Centre (CGFDC), founded in 1993, is directly under the authority of the Ministry of Agriculture and is responsible for international cooperation, technical promotion, and quality control of green food (Sanders 2006). The Centre became a member of the International Federation of Organic Agricultural Movements (IFOAM) in the same year in an attempt to gain international acceptance of green food, which proved to be naïve (Thiers 2005). To target international organic markets without abandoning what has been accomplished domestically, the centre formulated two types of green food standards in 1995: "A" and "AA" grades (Zong 2002); the latter was comparable with organic standards (Thiers 2002; Lin, Zhou, and Ma 2010). Most references to green food are mainly to the Grade A standard. Green food production has been predominantly conducted on state farms,[2] where land is farmed on a large scale. Based on the House-hold Responsibility System, most green food farming outside state farms has been undertaken by cooperatives rather than individual households. Earning a higher income is the primary incentive for most farmers (or cooperatives) converting to green certified production (Sanders 2006).

Chinese government agencies have played important roles in initiating and stimulating green food development. Nationwide, by 2011, 42 provincial and municipal branch agencies had been established to certify and manage green food at the local level. In addition, 38 quality inspection stations and 71 environmental monitoring branches had been established with authorization from the national office to ensure compliance with green food standards. Many farmers were initially encouraged by government officials at local levels to convert to green food production (Sanders 2006). This green food program has become a remarkably successful ecolabelling innovation because of its rapid growth rate in past decades, its similarities to the organic sector, and subsequent contributions to China's organic "revolution" (Giovannucci 2005; Mei, Jewison, and Greene 2006; Paull 2008).

Hazard-free food

In 2001, 11 years after the green food program was launched, the Ministry of Agriculture launched a Hazard-Free Food Action Plan to address the expanding food safety problems and high agrochemical contamination issues. Compared to green food standards, the hazard-free standards, also known as "pollution-free" quality standards, permit a wider range of agrochemicals. Genetically modified (GM) foods are allowed within the hazard-free standard. GM cotton and papaya production are permitted in China and are widely grown, but GM grain production is not (Wong and Chan 2016). GM ingredients are allowed in processed green and hazard-free food (China Green Food Development Center website 2018). Table 3.1 outlines the differences among green, organic, and hazard-free quality standards.

The rationale for launching a hazard-free food program was twofold: China needed to establish basic food standards for food sold through conventional value chains; and green and organic food quality standards were considered to be too stringent to be widely adopted in China given the severely polluted environment (i.e. the land couldn't get certified for organic or green food production) (Jia et al. 2002). Several officials mentioned the long-term plan of making the hazard-free certification standard a basic minimum requirement for agrofood production to ensure food safety.[3]

Organic agriculture

Organic agriculture is an international food quality standard, which was first introduced in China by the Dutch certifier SKAL in 1990 (Zong 2002). At that time, with a low awareness of organic agriculture and a limited demand in China's domestic market, certified organic agriculture was mainly for export. China implemented its own national organic standards in 2005, and introduced a national label for organic food sold domestically. Certified organic agriculture increased rapidly in the 2000s both in terms of the certified land area and the market value of certified organic products (Scoones 2008; Sternfeld 2009). Since that time, the Chinese domestic market for organic products has grown dramatically, largely attributed to the expanding population of middle and upper classes who have stronger purchasing power and more awareness of nutrition, health, and food safety issues (Xu 2008). To raise consumer confidence in organic certification and to better regulate organic food markets, a set of revised, more stringent national organic standards was introduced in 2012. The new standards have stricter labelling and traceability requirements on

Table 3.1 Comparison of organic agriculture, green food, and hazard-free food

	Organic agriculture (youji shipin)	Green food (lüse shipin)	Hazard-free food (wu gonghai shipin)
Year established	1994 (national standards passed in 2005)	1990	2001
Regulatory body	Jointly overseen by the MoA and the Ministry of Environmental Protection	Ministry of Agriculture	Ministry of Agriculture
Permits genetically modified organisms?	No	Yes	Yes
Permits synthetic fertilizer and pesticides?	No	Yes (only some kinds of chemical applications are permitted and amounts are regulated)	Yes (a wider range of agro-chemicals are allowed than for green food)
Residue testing	Yes	Yes	Yes
Initial force	Government and large agribusinesses for exports	Government and market	Government-initiated
Certifiers and certification costs	Third party certification; 20–40,000 CNY* (before new regulations in 2012)	Ministry of Agriculture's Green Food Development Centre; 10,000 CNY**	Ministry of Agriculture's Center for Agri-Food Quality and Safety; no fee
Traceability	Yes	No	No
Period of validity	One year	Three years	Three years

Source: Authors' compilation, adapted from Scott et al. (2014, p. 161).

Notes
*   Approximately US$3,000–6,000.
** Approximately US$1,500.

organic products. According to the new standard, all organic certified products on the market have been given a unique 17-digit tracing number (on the smallest package), which can be verified easily by consumers.[4]

In 2016, China was the fifth largest country in the world in terms of the land area for certified organic cultivation, with 2.28 million hectares and an organic market of US$6.6 billion (Willer and Lernoud 2017a, CNCA and CAU 2017). The growing trend of organic agriculture in terms of total certified organic land area from 2005 to 2015 is illustrated in Figure 3.1.

Chinese paths in promoting organic agriculture

Over the past few decades, organic sectors globally saw a transformation in ownership structures from a predominance of small independent farms to greater conventionalization, or a bifurcation of large and small actors. In the organic agriculture sector in China, in contrast, the ownership structure has undergone a transformation from being dominated by trading companies (in the form of contract farming arrangements) to the co-existence of various models. This transformation has happened alongside the expansion of the domestic organic market and the promotion of farmers' cooperatives as a development strategy (see Table 3.2).

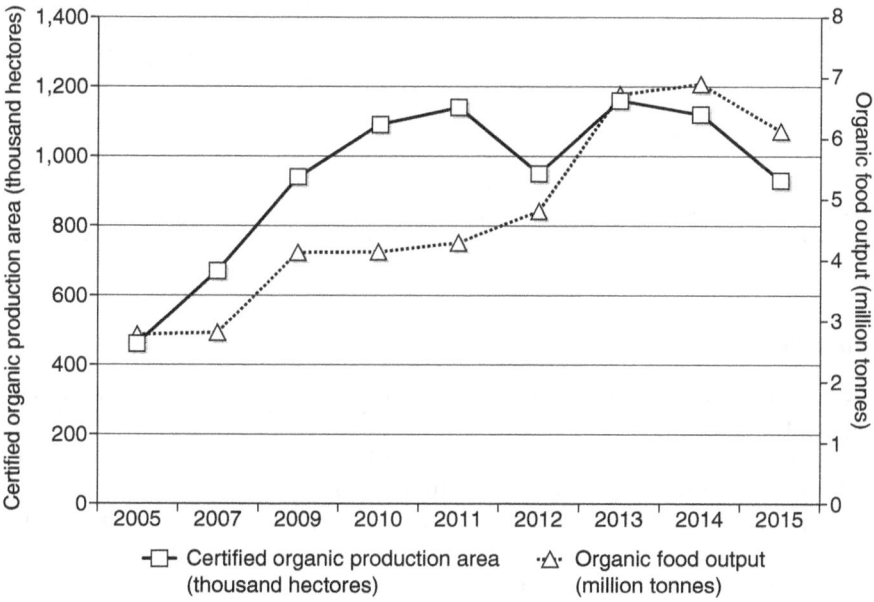

Figure 3.1 Development of certified organic agriculture in China, 2005 to 2015.

Source: CNCA & CAU 2017.

Note

No data available in 2008; Data in 2006 are debatable and thus not included.

Table 3.2 Characteristics of the emergence and subsequent development phases of China's organic agriculture sector

	Emergence of organic agriculture sector (1990s)	*Further development (since mid-2000s)*
Driving forces	Chinese government, in cooperation with Chinese and foreign trading companies	Agribusiness, farmer professional cooperatives, plus civil society (consumers and NGOs in alternative food networks)
Grower motivations	Agribusinesses: profit Small producers: guaranteed markets with a moderate price premium	Profits and some values-based initiatives (environmental and broader social values)
Market channels	Export markets	A combination of export and domestic markets; the latter includes direct marketing (farm to customer) and conventional marketing channels (e.g. via supermarkets)
Ownership structure	Contract farming with small-scale farmers	Various ownership structures: enterprises leasing land; contract farming; farmer professional cooperatives
Government roles	Intervention of local government agencies, including using administrative measures to force farmers to swap land and convert to organic agriculture; limited civil society involvement	Legislated standards, and various policy supports (e.g. subsidies, tax exemptions, loans, and facilitating land access)

Source: authors' fieldwork.

Strong government roles

The Chinese government, in cooperation with the private sector, has played a strong role in initiating and developing the organic agriculture sector since the early 1990s. China's organic agriculture sector was initiated and organized by local government officials and export-oriented trading companies (Thiers 2005). Realizing the potential to increase small farmers' incomes and boost local economic development, Chinese government agencies at different levels have provided various supports to promote organic production, using diverse administrative procedures, financial support, and other incentives (Scott et al. 2014; Qiao et al. 2018; Qiao, Martin, and Pan 2018).

The strong government roles in China's organic sector development can be explained by a number of factors. Although rural reforms have provided broad autonomy for farmers and allowed business enterprises to become important actors in the agriculture sector, local governments in China still remain the leading actors in rural economic development (Zhang 2012). Realizing the potential to alleviate rural poverty and support rural development, ecological and organic agriculture has been promoted by local governments in China, especially in wealthier provinces, since the 1990s. Local and provincial governments, often working closely with the private sector, have strong incentives to improve the economic performance in their jurisdiction in order to generate more tax revenues. To draw in businesses and investment to the agricultural sector in their region, local government agencies have provided generous subsidies, preferential loans, and tax exemptions. As a result of China's fiscal decentralization in the mid-1980s, local government revenues are mainly determined by the economic performance in their jurisdiction (Walder 1995). Moreover, the private sector is also keen to establish political connections in China for reasons such as accessing state-controlled resources (e.g. obtaining loans from banks and other state institutions) and overcoming legal and institutional failure and discrimination against private ownership (see Fan, Wong, and Zhang 2007; Li et al. 2008).

In addition to the roles of local government mentioned above, the central government has also played a role in organic sector development. This includes the institutionalization of organic and ecological agriculture through establishing production standards and certification procedures. Key Chinese government agencies that have been involved in promoting organic agriculture include the Ministry of Agriculture (MOA), the Ministry of Environmental Protection (MEP), the Ministry of Commence (MOC), the Development and Reform Committee, and the Ministry of Science and Technology (Qiao 2011). At the outset, major certification agencies in China (e.g. OFDC and COFCC) were affiliated with certain government departments. For example, OFDC was affiliated with and administered under the State Environmental Protection Agency (later became the MEP) when it was established in 1994. Many of the early certified organic farms were state farms, which resulted in conflicts of interest, as state entities acted as both regulators and competitors (Thiers 2002). In the 2000s, the certification agencies were gradually divorced from their affiliated departments and became third-party certification agencies.

At the provincial and local levels, government agencies have played various roles in promoting and facilitating the organic agriculture sector within their jurisdiction by using both political and economic resources. Organic agriculture has been viewed as a development strategy by the Chinese local government for the benefit of themselves and rural development. Based on our fieldwork and a review of secondary sources, we classified local government efforts in facilitating organic agriculture sector into the following categories (see also Scott et al. 2014):

- Administering standards and testing: agricultural bureaus[5] at the provincial, prefectural, and county levels are responsible for ensuring the enforcement of ecological and organic standards and testing agricultural products.[6]
- Providing or facilitating land access: helping agribusinesses get access to farmland for organic production either through direct provision of land (from government) or through negotiating land leasing with small-scale farmers to consolidate rural land for investors (Kledal and Sulitang 2007).
- Developing regional organic product development plans and establishing eco-agricultural zones, organic agriculture gardens, and demonstration bases: some local government officials are ambitious enough to promote organic production across a whole county, such as Wanzai organic county in Jiangxi province and Baoying organic country in Jiangsu province (see Qiao et al. 2018; Qiao, Martin, and Pan 2018).
- Financial support and subsidies, such as subsidies for building greenhouses and installing irrigation facilities, which are often not designed specifically for organic agriculture, and subsidies for organic fertilizer and bio-pesticides (see also Bennett 2009). For example, the Shanghai municipal government invested US$30 million in organic fertilizer subsidies between 2004 and 2009. There are also specific subsidies for organic agriculture in some jurisdictions, including subsidizing organic certification fees, pest repellent lamps and pest traps; offering low or no interest loans; and tax exemptions for agribusiness and cooperatives.
- Technical support, such as training workshops about organic production techniques, organizing visits to other organic farms or production bases, or bringing in researchers to provide advice.
- Marketing support: assisting with branding, organizing expos, and other forms of product promotion; institutional procurement, such as purchasing organic food as a perk for government employees; promoting ecological agritourism; helping to host various organic and green food expos, such as the BioFach China and the China Green Food Expo.
- Facilitating vertical integration to reduce transaction costs associated with contracting a large number of small-scale farms: helping establish farmers' cooperatives, since trading and processing companies prefer to contract with cooperatives rather than numerous small-scale farmers to transition to organic production.
- Commemoration: various awards provided by various levels of government in recognition of good performance of organic farms, such as the "organic agriculture demonstration model" at national, provincial, or city levels.
- Improving public awareness of ecological agriculture and organic food. For example, the China Environment and Sustainable Development

Reference and Research Center (CESDRRC, a sub-unit of the State Environmental Protection Agency) started a monthly newsletter, *Organic Trends*. This centre also worked with NGOs (e.g. Friends of Nature and Global Village Beijing) to set up a webpage, Green Choice, as a platform to discuss food safety and organic food.

* Other roles: state-owned enterprises facilitating their suppliers' conversion to organic production (Sanders and Xiao 2010); encouragement to convert to organic agriculture (Sanders 2006).

Thus, realizing the potential to boost the local economy and protect the natural environment, local governments have provided diverse supports to encourage ecological and organic agriculture development within their jurisdictions. There are a few caveats, however.

First, although ecological and organic agriculture has been viewed as an important strategy in alleviating rural poverty, enhancing food safety, and developing sustainable agriculture, our research found that the central government does not play an active role in this sector in practice besides issuing policies to encourage the development and setting up of demonstration sites. Government roles at both local and central levels in educating the public and improving awareness of organic and ecological agriculture is still lacking, and consumers are often confused about the differences between the various food quality standards (organic, green, and hazard-free food).

Second, while the range of local government support may be impressive, there are substantial variations by region due to differing financial capabilities of provincial and local governments. Government agencies in wealthier areas more often promote organic agriculture through direct financial support and subsidies, while tax exemption and other kinds of policy support are more common in poorer areas.[7] Strong local government involvement in the rural economy, however, has also led to jurisdictional competition in which state entities—local government agencies, subunits of ministries, and research institutions—invest both capital and political authority to compete to build a business environment favourable to private capital investment to organic agriculture (Qian and Roland 1998).

Third, compared to green and hazard-free food, organic agriculture has received relatively less support from local and provincial governments. In fact, most of the government support that organic farms receive is actually support for agriculture in general, rather than support designed specifically for organic farms. This may be due to the skepticism among government officials, and some organic farm operators, regarding the productivity of organic agriculture, and the concerns that expansion of organic production could lead to insufficient food supplies to feed China's population.[8] In consideration of the already serious environmental situation in China, the less stringent standards of green and hazard-free are often viewed as more suitable to be widely adopted because most farms would not pass organic certification.[9]

Expanding domestic markets for ecolabelled food products

Organic food in China consists mostly of fresh produce (such as fruit and vegetables), dried or frozen field crops (such as grains, corn and beans), and tea. Most certified organic products in China are available in large urban centres, with the majority of them being consumed by upper- and middle-class individuals, particularly families with pregnant women, young children, or people with health problems (Sheng et al. 2009; Shi et al. 2011a; Thøgersen and Zhou 2012). In 2006, official data indicated that, for the first time, the domestic sale value exceeded the export value, although exports also continued to increase, from US$150 to US$350 between 2004 and 2005 (Sternfeld 2009). The value of domestic sales organic food was approximately US$1.7 billion in 2009, which was almost four times of the value of exported organic products (US$464 million).[10] In 2013, 90 per cent of certified organic products grown in China were sold and consumed domestically, with only 10 per cent for export.[11] Consumer trust in food safety and quality dramatically decreased in China in the 2000s, mainly because of the growing ecological awareness and numerous food scandals exposed by the mass media, including the tainted infant formula scandal and recycled gutter oil used in cooking (Chen 2013; Mol 2014). As Shi Yan, the founder of the first CSA farm in China, explained,

> Certified organic agriculture in China has been export-oriented since its emergence in 1990s. But until a series of food safety scandals, especially the tainted infant formula scandal, were reported by the media in China in the mid-2000s, people started to pay high attention to food safety. And it is at this time that people started to purchase organic food and the Chinese domestic market for organic agriculture has been grown rapidly since then.
>
> (Interview with Shi Yan at Little Donkey Farm, Beijing, April 11, 2012)

The consumption of organic food in China has been motivated mainly by individual health concerns and strongly affected by economic factors like household income and organic food prices, rather than the environmental and other social and ethical values in purchasing and consuming organic products (Yin et al. 2010; see also Chapter 5).

Besides the rising health concerns, the burgeoning middle classes with strong purchasing power is another major factor contributing to the rapid growth of China's domestic organic market. Social and economic reforms have significantly contributed to China's growing economy and a growing middle and upper-middle class. Compared with the lower class, whose concerns related more to food security, middle-class consumers have strong concerns around food safety and quality and are willing to pay extra for better-tasting and higher-quality food (Banerjee and Duflo 2008). The large Chinese middle-class population and their growing awareness of nutrition, health, and food safety account for the expanding Chinese domestic organic market (Xu 2008).

Although the Chinese domestic market has grown rapidly since the 2000s, it still faces many barriers to better meet consumer demand. These barriers include low availability of organic products, high prices, low trust in certification and ecolabelling, and limited knowledge among consumers about ecological agriculture and ecolabelled products.[12] Yin et al. (2010) found that the most important reasons for not purchasing organic food were high prices and limited availability on the market. In addition, low trust and limited knowledge about ecological agriculture and ecolabelled products increase consumer uncertainty, which will have negative impacts on the consumers' willingness to buy ecolabelled products (Thøgersen and Zhou 2012). Moreover, the mass media has also exposed various organic food scandals. These food safety and food quality scandals have hit the organic food industry and have had negative impacts on consumers' purchases of organic food (Yin et al. 2010). To overcome these barriers, some solutions have been proposed, including diversifying the marketing channels to cut down the market price and improve market availability, lowering the premium prices, enhancing the monitoring system, and reducing consumer uncertainty by means of consumer education and campaigns (Yin et al. 2010; Thøgersen and Zhou 2012).

Co-existence of diverse ownership structure in China's organic agriculture sector

The growing domestic organic market has provided opportunities for private companies and farmers' professional cooperatives to expand organic production through direct marketing strategies. To capture the tremendous growth potential and reap high price premiums, private investors from various sectors (real estate, IT, etc.) have shown strong interest and made substantial investments in China's organic agriculture sector (Yuan 2011). For example, Tony's Farm, one of the earliest and most well-known organic farms in China, has invested over 250 million CNY (equal to US$41 million) since its founding in 2005, with the funding coming from their own enterprise, local government, and venture capital funds. More recently, some large agribusiness companies, which engaged in contract farming and exported products to the global market, have begun to develop their sales domestically.[13] The expanding domestic organic market contributes significantly to the viability of diverse ownership structures, scales of production, and types of market channels in China's organic sector. Based on review of literature and our fieldwork findings, we identified three main models of ownership structure of organic farms in China's organic sector: the contract farming model, the private company land leasing model, and the independent farmer professional cooperative (FPC) model (see Table 3.3).

Table 3.3 Characteristics of three types of ownership structure of organic farms in China's organic sector

	Contract farming model (company + cooperative)	Private company land leasing model	Independent farmer professional cooperatives (FPCs) model
Management structure	Processing/trading companies + FPCs + small-scale farms	Company hires farm workers	FPC members
Roles	Company arranges certification, technical assistance and guaranteed prices; smallholders (via FPCs) grow and sell the agricultural products	Company hires farm labourers to work on the farm	FPC arranges certification, production, processing and marketing
Source of land	Rural households' land	Leased from farmers or village government	Rural households' land

Source: authors' compilation.

Contract farming

The contract farming model we discuss here refers to an arrangement in which food companies signed production contracts with farmers' cooperatives or a group of small farmers to produce certain food products following specific requirements. Farmers often had to meet delivery requirements of the company and the company often provided various supports such as coordinated purchase of farming inputs and other technical support, and agreed to buy the output at a preset price. Traditionally viewed as an institutional solution to a number of barriers related to access to credit, insurance, and technical farming advice encountered by small-scale farmers (Key and Runsten 1999), contract farming is typically driven by large-scale food processing/trading companies to ensure that a steady supply of agricultural commodities meets certain quality standards.

Realizing the potential to link numerous small-scale farmers to a wider market and achieve agricultural industrialization, the contract farming model was promoted strongly by the Chinese government with subsidies, preferential loans, and tax exemptions (Huang 2011; Huang et al. 2012). Many large agribusiness enterprises, most of which have the status of "dragon-head enterprises" in China (see discussion in Chapter 2), uses contract farming as a business strategy to reduce their risk and uncertainty and to achieve greater economic returns. Before 2012, these trading and processing companies more commonly chose to sign contracts with farmers' cooperatives rather than creating relatively large-scale farms themselves, as was typical in many other countries (e.g. see Singh 2002). Since 2012, however, we have seen a growing trend of agrifood companies setting up ecological agriculture farms by leasing land from farmers and hiring farm workers.

In the contract farming model, local officials at the village level often acted as a broker between the company and the farmers by signing a long-term contract of organic production and purchasing. By contracting with farmers' cooperatives, agribusiness enterprises can also help strengthen their own positions by ensuring favourable regulatory treatment from the government and access to productive resources. Farmers' cooperatives, as we discuss in more detail in Chapter 4, were viewed by the Chinese government as a suitable institutional unit for organizing, monitoring, controlling, and taxing agricultural production. In this model, processing and trading companies were mainly responsible for processing and marketing, and sometimes also supplying inputs and providing necessary technical support. The farmers' cooperatives were mainly responsible for agricultural production. One such example is the Tai'an Asian Food Company (TAAFC) in Shandong province. In 2006, the TAAFC entered agreements with 1,300 households in 17 villages to convert 534 ha to organic vegetable production.[14] Besides providing a guaranteed market at favourable prices, this company also arranged organic certification and provided on-site technical training and monitoring through hiring a technical expert to reside in each village. Another example of this model is the state-owned Maotai Company, which facilitated the conversion to organic production among its suppliers of raw materials, primarily sorghum and wheat (Sanders and Xiao 2010).

Private company land leasing

Under the private company land leasing model, entrepreneurial farmers and investors managed organic production by leasing land from small farmers and hiring farmers to work on the company-controlled land. This model was originally developed in coastal provinces where farmers can easily find jobs in non-agricultural sectors and the farmland has been idle or underutilized for many years.[15] It has become increasingly popular in many other regions in recent years as the government has loosened its restrictions on farmland transfer to encourage "moderate-scale" operation of agriculture. To better utilize farmland, some villages, townships, or counties lease land from local farmers and manage it in a unified way, investing in roads, water supply, and/or greenhouses, in order to attract external investors. In some such areas, a strong preference has been given to investors interested in organic and ecological production. The county districts of Shanghai have been major actors in promoting this model of organic production (also see Kledal and Sulitang 2007).[16] These types of organic farm were varied in scale of operation. They sometimes choose to target the domestic market given their well-established market networks and geographical proximity.[17] Products are sold via various marketing channels, including supermarkets and direct marketing channels.

Farmer professional cooperatives (FPCs)

To overcome the limitations of small-scale production, some small-scale farmers chose to establish or join farmers' cooperatives in which they pool resources (other than land) and market their produce collectively (Zong 2002; Thiers 2005; Mei, Jewison, and Greene 2006). By working together, small-scale farmers can reduce transaction costs and increase their bargaining power in the supply chain (Bosc et al. 2002; Stringer, Sang, and Croppenstedt 2009). Unlike farmers' cooperatives under the contract farming model, in which cooperatives are subordinate to enterprises, independent FPCs were established and managed by farmers themselves and FPCs represent members' interests (Huang et al. 2012). Although no data are available about the number of independent farmers' cooperatives conducting ecological and organic agriculture in China, by the end of November 2017 there were more than 1.99 million FPCs nationwide in China, 76 times more than in 2007, when the Farmers' Professional Cooperative Law was passed in China (Xinhua Net 2017), as we discuss in the next chapter. However, there has been strong suspicion about the authenticity of these farmers' cooperatives. Some analysts estimate that 80 to 95 per cent of these are fake either because they exist on paper only or because they are actually food companies trying to use the label of cooperative to qualify for government subsidies (see Yan and Chen 2015).

Owing to the costly certification fees, the revised national organic standards that were introduced in 2012 were expected to have significant impacts on the ownership structure of China's organic production, since many organic enterprises would choose to drop certification or to decrease the number of certified products to reduce costs (Scott et al. 2014). Official statistics revealed that one-third of China's 7,000 organic enterprises left the organic market in the first few months of the implementation of the new standards (Yu 2012). Because of the stronger financial and managerial capacity, large agribusiness enterprises were expected to be better positioned than others under the more stringent standards and higher costs for organic certification (also see Horowitz 2012).[18] Few household-scale organic farm operations have been able to gain certification on their own due to a variety of reasons: the high cost of certification; farming households are generally small-scale with limited resources; and the difficulties of seeking external support and accessing value-added markets on their own (Zong 2002).[19] Some previously certified small-scale organic enterprises opted to continue following organic management practices but without getting certified, and explaining this reasoning to their customers. Rather than relying on certification, these farmers reconnect with customers and sell their products directly to them through various channels.[20]

Conclusion

The certified organic agriculture sector in China has undergone significant changes over its short period of development, mainly in terms of the ownership structure of organic farms and their marketing channels. We argue that organic agriculture in China can be characterized in terms of three major dimensions. First, China's path to promote ecological agriculture is characterized as a top-down process. The government has played strong roles in its development. Second, although China's organic agriculture was promoted initially for export markets, China has witnessed a growing domestic market for "safe" and pesticide-free food since the 2000s. Third, the unique land tenure system based on the Household Responsibility System and the growing domestic markets shape the co-existence of diverse ownership structure in China's organic agriculture sector.

As in many other developing countries, organic agriculture in China was initiated in the 1990s for export, and there was limited concern among Chinese consumers about environmental conservation or social justice. Our study found that, along with the development of organic agriculture, especially the growing domestic organic market in the 2000s, the ownership structure in China's organic sector has diversified with the co-existence of contract farming, private companies land leasing, and independent farmers' professional cooperative models. After more than a decade of development, the private companies land leasing model is more popular than others. The Chinese government, in cooperation with the private sector, has played a strong role in developing the organic sector. The government provides various supports to the sector, yet it favours large farms and excludes small farmers from accessing these supports.

Besides these certified organic farms, a small number of values-based initiatives have started to emerge, which pay more attention to the broader values of organic production and highlight the importance of direct interactions between consumers and farmers to rebuild relationships of trust. They often adopt direct marketing strategies, such as CSA, home delivery, buying clubs, and farmers' markets, to sell their products. Consumers' trust is enhanced through direct interactions in these initiatives rather than through official organic labels. These are often bottom-up initiatives that receive much less support from the government but involve more participation of community and social organizations. Chapters 5 through 8 examine these initiatives in detail from various dimensions.

The study of China's organic production reveals a trend of diversification in ownership structures in organic production with an emergence of values-based initiatives. This study contributes to our understanding of the complexity and diversity of the development path of organic agriculture in China. The findings of this research give us a glimpse at the development patterns of organic agriculture beyond the West, but also serve as a mirror to reflect the possible trajectories of organic agriculture in emerging economies that need further investigation. Many other emerging economies, such as India and Brazil, face situations in developing organic agriculture similar to that found in China. In these countries, organic agriculture has been driven mainly by export markets via the contract farming model; and domestic organic markets have been emerging over the past decade, providing some more opportunities for small-scale farmers to engage in and benefit from the organic agriculture sector (Egelyng et al. 2010; Menon, Sema, and Partap 2010; Blanc and Kledal 2012; Osswald and Menon 2013).

Notes

1 In 2018, the Ministry of Agriculture was renamed the Ministry of Agriculture and Rural Affairs.
2 In China, state farms are owned by the state and operated directly under the MOA rather than allocated to individual rural households. There are approximately 2,000 state farms, operating in 30 provinces and covering 39 million hectares of land, or 4 per cent of China's total rural land. On these farms, all assets, including land, buildings, machines, and farm animals, are owned by the state (Zhang 2010).
3 Personal interviews with government officials and staff from the China Organic Food Certification Center (COFCC) in Beijing, multiple dates, in 2011.
4 Interview (via phone) with Zhou Zejiang, Vice President of IFOAM Asia, Nanjing, August 28, 2014.
5 CNCA is now in charge of organic certification in China, but this department does not have a vertical hierarchy with subordinate bureaus and offices at the local levels to administrate the implementation of organic standards. So, the subordinate bureaus and offices of the then MOA are responsible for the administration and product testing.
6 Organic produce samples need to be tested at an agency with legitimate qualifications entrusted by the organic certification agency.
7 Interviews with government officials at the provincial and county levels at Zhejiang, Jiangsu, Anhui, Shanghai, and Beijing in 2010 and 2011.
8 Interview with an agricultural scientist, Beijing, April 13, 2012.
9 Interviews with government officials and farmers, Jiangsu, Anhui, Shandong, and Hainan, various dates, 2010–2011.
10 Interview with Gao Xiuwen, Assistant Director, Certification Management Department, China Organic Food Certification Centre, Beijing, April 10, 2012.
11 Interview with Zhou Zejiang, Vice President of IFOAM Asia, Nanjing, November 22, 2014.
12 Interview with various stakeholders in 2010 and 2011 in China.
13 Interview with the CEO of Tai'an Asian Food Company (TAAFC), Tai'an, Shandong province, March 17, 2011.
14 Personal interview with the CEO of Tai'an Asian Food Company, March 17, 2011 in Tai'an, Shandong province.

15 Personal interviews with government officials at Shanghai and Nanjing and organic certifiers at Nanjing, 2010 and 2011.
16 Personal interview with government official in Shanghai, March 20, 2011.
17 Interview with several CEOs/managers of farms operating with the private company leasing land model in 2010–2011.
18 Interview with Zhou Zejiang, Vice President of IFOAM Asia, Nanjing, March 11, 2013.
19 These factors were also confirmed by staff from organic certification agencies (e.g. the OFDC, COFCC and Eco-Cert) in Jiangsu and Beijing, various dates, 2010–2011.
20 Interview with Zhou Zejiang, Vice President of IFOAM Asia, Nanjing, March 11, 2013.

Bibliography

Banerjee, A and Duflo, E 2008, "What is middle class about the middle classes around the world?" *Journal of Economic Perspective*, vol. 22, no. 2, pp. 3–28.

Bennett, MT 2009, *Markets for Ecosystem Services in China: An Exploration of China's "Eco-compensation" and Other Market-Based Environmental Policies.* A Report from Phase I Work on an Inventory of Initiatives for Payments and Markets for Ecosystem Services in China. Washington, DC (Forest Trends). Accessed at http://forest-trends.org/publication_details.php?publicationID=2317 on June 9, 2011.

Blanc, J and Kledal, P 2012, "The Brazilian organic food sector: prospects and constraints of facilitating the inclusion of smallholders," *Journal of Rural Studies*, vol. 28, no. 1, pp. 142–154.

Bosc, PM, Eychenne, D, Hussein, K, Losch, B, Mercoiret, MR, Rondot, P, and Mackintosh-Walker, S 2002, *The Role of Rural Producer Organizations in the World Bank Rural Development Strategy.* The World Bank Rural Development Family: Rural Development Strategy, background paper. World Bank, Washington, DC.

Browne, A, Harris, P, Hofny-Collins, A, Pasiecznik, N, and Wallace, R 2000, "Organic production and ethical trade: definition, practice and links," *Food Policy*, vol. 25, no. 1, pp. 69–89.

Chen, R 2013, "Discovering distinctive east Asian STS: An introduction," *East Asian Science, Technology and Society: An International Journal*, vol. 6, no. 4, pp. 441–443.

China Green Food Development Center website, 2018, www.greenfood.org.cn/ywzn/lssp/jsbz/xxlsspbztx/index_1.htm.

CNCA (Certification and Accreditation Administration of the PRC) and CAU (China Agriculture University) 2017, *The Development of Organic Products Certification and Industry in China (2016)*, China Quality Inspection Publishing House and Standards Press of China, Beijing (in Chinese).

Egelyng, H, EI-Araby, A, Kledal, P, and Hermansen, J 2010, *Certified Organic in a North-South and South-South Perspective.* Paper presented at the International Symposium "Governing through Standards," Copenhagen, Denmark, February 24–26.

Fan, J, Wong, TJ, and Zhang, T 2007, "Politically-connected CEOs, corporate governance and post-IPO performance of China's newly partially privatized firms," *Journal of Financial Economics*, vol. 84, no. 2, pp. 330–357.

Giovannucci, D 2005, *Organic Agriculture and Poverty Reduction in Asia: China and India Focus*, Report No. 1664. International Fund for Agricultural Development (IFAD), Rome.

Gong, H, Meng, D, Li, X, and Zhu, F 2013, "Soil degradation and food security coupled with global climate change in northeastern China," *Chinese Geographical Science*, vol. 23, no. 5, pp. 562–573.

Horowitz, S 2012, *Building Trust or Burning Bridges? Market Implications of China's New Organic Certification Standards.* Unpublished.

Huang, J, Wang, X, and Qiu, H 2012, *Small-Scale Farmers in China in the Face of Modernisation and Globalisation.* IIED/HIVOS, London/The Hague.

Huang, P 2011, "China's new-age small farms and their vertical integration: Agribusiness or co-ops?" *Modern China*, vol. 37, no. 2, pp. 107–134.

IPES-Food 2016, *From Uniformity to Diversity: A Paradigm Shift from Industrial Agriculture to Diversified Agroecological Systems.* Accessed at www.ipes-food.org/images/Reports/UniformityToDiversity_FullReport.pdf on May 11, 2017.

Jia, NX, Liu, HF, Wang, XP, and Liu, Y 2002, "Discussion on the development of organic food, green food and hazard free food," *Journal of China Agricultural Resources and Regional Planning*, vol. 23, no. 5, pp. 60–62 (in Chinese).

Kahrl, F, Li, Y, Su, Y, Tennigkeit, T, Wilkes, A, and Xu, J 2010, "Greenhouse gas emissions from nitrogen fertilizer use in China," *Environmental Science and Policy*, vol. 13, no. 8, pp. 688–694.

Key, N and Runsten, D 1999, "Contract farming, smallholders, and rural development in Latin American: the organization of agro-processing firms and the scale of outgrower production," *World Development*, vol. 27, no. 2, pp. 381–156.

Kledal, PR and Sulitang, T 2007, *The Organization of Organic Vegetable Supply Chains in China—Flexible Property Rights and Different Regimes of Smallholder Inclusion*, 106th EAAE seminar, October 25–27, Montpellier, France.

Li, H, Meng, L, Wang, Q, and Zhou, L 2008, "Political connections, financing and firm performance: Evidence from Chinese private firms," *Journal of Development Economics*, vol. 87, no. 2, pp. 283–299.

Li, H, Zeng, EY, and You, J 2014, "Mitigating pesticide pollution in China requires law enforcement, farmer training, and technological innovation," *Environmental Toxicology and Chemistry*, vol. 33, no. 5, pp. 963–971.

Li, J 2013, "China gears up to tackle tainted water; government is set to spend 500 million renminbi to clean up groundwater polluted by industry and agriculture," *Nature*, vol. 499, no. 7456, pp. 14.

Lin, L, Zhou, D, and Ma, C 2010, "Green food industry in China: Development, problems and policies," *Renewable Agriculture and Food Systems*, vol. 25, no. 1, pp. 69–80.

Liu, R, Pieniak, Z, and Verbeke, W 2013, "Consumers' attitudes and behaviour towards safe food in China: A review," *Food Control*, vol. 33, no. 1, pp. 93–104.

Mei, Y, Jewison, M, and Greene, C 2006, *Organic Products Market in China.* USDA Foreign Agricultural Service, GAIN Report, Beijing, China.

Menon, M, Sema, A, and Partap, T 2010, "India organic pathway: Strategies and experiences." In: Partap, T and Saeed, M (eds.), *Organic Agriculture and Agribusiness: Innovation and Fundamentals.* The Asian Productivity Organization Hirakawacho, Chiyoda-ku, Tokyo, pp. 75–86.

Mol, A 2014, "Governing China's food quality through transparency: A review," *Food Control*, vol. 43, pp. 49–56.

Organic Monitor 2017, *The Global Market for Organic Food and Drink*. Accessed at www.organicmonitor.com.

Osswald, N and Menon, M 2013, *Organic Food Marketing in Urban Centres of India*. ICCOA, Bangalore.

Paull, J 2008, *The Greening of China's Food: Green Food, Organic Food, and Eco-labelling*. Paper presented at the Sustainable Consumption and Alternative Agri-Food Systems Conference, Liege University, Arolon, Belgium.

Qiao, Y 2011, "Organic agriculture development in China." In: Willer, H and Kilcher, L (eds.) 2011, *The World of Organic Agriculture. Statistics and Emerging Trends 2011*. FiBL-IFOAM Report. IFOAM, Bonn and FiBL, Frick, pp. 132–136.

Qian, Y and Roland, G 1998, "Federalism and the soft budget constraint," *American Economic Review*, vol. 88, no. 5, pp. 1143–1162.

Qiao, Y, Martin, F, Cook, S, He, X, Halberg, N, Scott, S, and Pan, X 2018, "Certified organic agriculture as an alternative livelihood strategy for small-scale farmers in China: A case study in Wanzai county, Jiangxi province," *Ecological Economics*, no. 145, pp. 301–307.

Qiao, Y, Martin, F, and Pan, X 2018, "The role of local governments in promoting organic agriculture: From driver to guide—Case study of Wanzai county, Jiangxi province, China," *Canadian Journal of Development Studies*.

Sanders, R 2000, *Prospects for Sustainable Development in the Chinese Countryside: The Political Economy of Chinese Ecological Agriculture*. Ashgate, Brookfield, VT, USA.

Sanders, R 2006, "A market road to sustainable agriculture? Ecological agriculture, green food and organic agriculture," *Development and Change*, vol. 37, no. 1, pp. 201–226.

Sanders, R and Xiao, X 2010, "The sustainability of organic agriculture in developing countries: Lessons from China," *International Journal of Environmental Cultural, Economic and Social Sustainability*, vol. 6, no. 6, pp. 233–243.

Scoones, S 2008, *Organic Agriculture in China—Current Situation and Challenges*. EU-China Trade Project.

Scott, S, Si, Z, Schumilas, T, and Chen, A 2014, "Contradictions in state- and civil society-driven developments in China's ecological agriculture sector," *Food Policy*, vol. 45, no. 2, pp. 158–166.

Sheng, J, Shen, L, Qiao, Y, Yu, M, and Fan, B 2009, "Market trends and accreditation systems for organic food in China," *Trends in Food Science and Technology*, vol. 20, no. 9, pp. 396–401.

Shi, T 2002, "Ecological agriculture in China: Bridging the gap between rhetoric and practice of sustainability," *Ecological Economics*, vol. 42, no. 3, pp. 359–368.

Shi, Y, Cheng, C, Lei, P, Wen, T, and Merrifield, C 2011, "Safe food, green food, good food: Chinese community supported agriculture and the rising middle class," *International Journal of Agricultural Sustainability*, vol. 9, no. 4, pp. 551–558.

Singh, S 2002, "Contracting out solutions: Political economy of contract farming in the Indian Punjab," *World Development*, vol. 30, no. 9, pp. 1621–1638.

Sternfeld, E 2009, "'Organic food' made in China," *EU–China Civil Society Forum Hintergrundinformationen*, vol. 10, pp. 1–12.

Stringer, R, Sang, N, and Croppenstedt, A 2009, "Producers, processors, and procurement decisions: The case of vegetable supply chains in China," *World Development*, vol. 37, no. 11, pp. 1773–1780.

Thiers, P 2002, "From grassroots movement to state-coordinated market strategy: The transformation of organic agriculture in China," *Environment and Planning C: Government and Policy*, vol. 20, no. 3, pp. 357–373.

Thiers, P 2005, "Using global organic markets to pay for ecologically based agricultural development in China," *Agriculture and Human Values*, vol. 22, no. 1, pp. 3–15.

Thøgersen, J and Zhou, Y 2012, "Chinese consumers' adoption of a 'green' innovation—the case of organic food," *Journal of Marketing Management*, vol. 28, no. 3–4, pp. 313–333.

Walder, A 1995, "Local governments as industrial firms: An organizational analysis of China' transitional economy," *American Journal of Sociology*, vol. 101, no. 2, pp. 263–301.

Wang, Z Q, Liu, B Y, Wang, X Y, Gao, X F and Liu, G 2009, "Erosion effect on the productivity of black soil in northeast China," *Science in China, Series D: Earth Sciences*, vol. 52, no. 7, pp. 1005–1021.

Willer, H and Lernoud, J 2017a, *The World of Organic Agriculture Statistics and Emerging Trends 2017*, FiBL-IFOAM Report.

Willer, H and Lernoud, J 2017b, *Organic Agriculture Worldwide 2017: Current Statistics*. Presentation at BIOFACH 2017, Nuremberg, February 15, 2017. Accessed at http://orgprints.org/31197/1/willer-lernoud-2017-global-data-biofach.pdf on May 1, 2018.

Wong, AYT and Chan, AWK 2016, "Genetically modified foods in China and the United States: A primer of regulation and intellectual property protection," *Food Science and Human Wealth*, vol. 5, no. 3, pp. 124–140.

Xinhua Net. 2017, "The newly amended Farmers' Professional Cooperative Law clarifies the legal status of farmers' professional cooperative unions." Accessed at www.xinhuanet.com/fortune/2017–12/27/c_1122176528.htm on May 2, 2018 (in Chinese).

Xu, F 2008, *Organic Products Shanghai Organic Retail Market Profile*. USDA Foreign Agricultural Service GAIN Report.

Yan, H and Chen, Y 2015, "Agrarian capitalization without capitalism? Capitalist dynamics from above and below in China," *Journal of Agrarian Change*, vol. 15, no. 3, pp. 366–391.

Yang, M 2012, "The damaging truth about Chinese fertiliser and pesticide use," *China Dialogue*. Accessed at www.chinadialogue.net/article/show/single/en/5153-The-damaging-truth-about-Chinese-fertiliser-and-pesticide-use on May 1, 2017.

Yin, S, Wu, L, Du, L, and Chen, M 2010, "Consumers' purchase intention of organic food in China," *Journal of the Science of Food and Agriculture*, vol. 90, no. 8, pp. 1361–1367.

Yu, C 2012, "Three months after new certification policy, organic supply shrinks: The difficulty of getting fresh organic vegetables on the tables of the masses," *Youth Times* (in Chinese). Accessed at www.qnsb.com/fzepaper/site1/qnsb/html/2012–11/04/content_398346.htm on May 2, 2018.

Yuan, Y 2011, "China's tycoons go farming," *China Dialogue*, November 10, 2011. Accessed at www.chinadialogue.net/article/show/single/en/4622-China-s-tycoons-go-farming on April 30, 2018.

Zhang, C, Guanming, S, Shen, J, and Hu, RF 2015, "Productivity effect and overuse of pesticide in crop production in China," *Journal of Integrative Agriculture*, vol. 14, no. 9, pp. 1903–1910.

Zhang, P 2009, "Rural interregional inequality and off-farm employment in China." In: Deng, Z (ed.), *China's Economy: Rural Reform and Agricultural Development*, World Scientific, Singapore.

Zhang, QF 2010, "Reforming China's state-owned farms: State farms in agrarian transition," Paper presented at the Asian Rural Sociology Association International Conference, Legazpi City, Philippines. Accessed at: https://ink.library.smu.edu.sg/soss_research/1089.

Zhang, QF 2012, "The political economy of contract farming in China's agrarian transition," *Journal of Agrarian Change*, vol. 12, no. 4, pp. 460–483.

Zhao, J, Luo, Q, Deng, H, and Yan, Y 2008, "Opportunities and challenges of sustainable agricultural development in China," *Philosophical Transactions of the Royal Society*, vol. 363, pp. 893–904.

Zhu, Z and Chen, D 2002, "Nitrogen fertilizer use in China—contributions to food production, impacts on the environment and best management strategies," *Nutrient Cycling in Agroecosystems*, vol. 63, pp. 117–127.

Zong, H 2002, "The China national study." In: *Organic Agriculture and Rural Poverty Alleviation, Potential and Best Practices in Asia*. United Nations Economic and Social Commission for Asia and Pacific (ESCAP), Bangkok, pp. 52–72.

4 The farmers' cooperative model in China's ecological agriculture sector

Aijuan Chen and Steffanie Scott

Introduction

Chapter 3 explained the major ownership structures of ecological farms, one of which was farmers' professional cooperatives (FPCs), which have received considerable attention due to their potential, dynamic growth, and management and development challenges. In this chapter, we provide an overview of the unique contribution of FPCs to sustainable rural development in China, and the implications for the evolution of the organic sector. FPCs have grown rapidly in rural China since the late 2000s. They have become an important institutional instrument for the Chinese government in attempting to achieve the vertical integration of agricultural production, processing and marketing.

Internationally, farmers' cooperatives tend to be more inclusive of the most resource-poor small farmers than contract farming models (Kruijssen, Keizer, and Giuliani 2009). In China, FPCs have been heralded as alternatives to large agribusiness companies by integrating small-scale farms with processing and marketing. Huang (2011) predicts that FPCs could outcompete agribusiness if they were given the same state subsidies and privileges. Others suggest that FPCs would be transformed into capitalist enterprises if farmers do not sustain strong anti-capitalist political mobilization (Hale 2013; Lammer 2012).

Gürel (2014) found that many FPCs in contemporary China are company-like cooperatives that are similar to agribusiness in terms of their "shareholding and decision making structures and the production relations they facilitate" (p. 69). Indeed, it has been estimated that 85–90 per cent of farmers' cooperatives in China are fake, either existing only on paper or established by private food enterprises posing as cooperatives, in order to acquire government subsidies and other support earmarked for FPCs (Yan and Chen 2015). These critiques tend to apply to cooperatives established by enterprises. Rather than continuing the debate on "real" and "fake" cooperatives here, we argue that FPCs—particularly the subset of cooperatives that are not merely extensions of agroenterprises—have the potential to make significant social, economic, and environmental contributions to rural development in China (Song et al. 2014; Yang et al. 2017). This can happen by applying the "deepening–broadening–regrounding" framework proposed by van der Ploeg, Long, and Banks (2002). In this vein, we sought to analyze how innovative and entrepreneurial strategies are pursued, what roles are played by the Chinese government in the establishment and operation of FPCs, and what roles are played by different farm members and their participation in decision-making and profit-sharing.

In this chapter, we present our research framework for analyzing the convergence of farmers' cooperatives for rural development. We introduce the methods used for data collection, and then provide an overview of FPC development in China. Next, we introduce three cases of FPCs and highlight the government's role in promoting FPCs. Finally, we analyze the contributions of FPCs and the challenges they face.

The cases examined for this study consisted of three cooperatives engaged in ecological and organic agriculture. Each of these three— Daizhuang Organic Farmers' Professional Cooperative in Jiangsu province, Tonglu Peach FPC in Zhejiang Province, and Yuexi Organic Kiwifruit FPC in Anhui province—represents an FPC that was initiated and established by different types of internal or external players. The cooperatives that we selected reflect the following three criteria: (1) They all follow the principles stated in the Farmers' Professional Cooperative Law, although all three FPCs existed before the law was enacted; (2) the cooperatives were initiated and established differently—by large-scale farms, by agroindustries, and by other external actors (such as researchers, government agents, foreign donors, and NGOs); and (3) they all adopt "alternative" farming strategies and have been relatively successful economically.

In order to demonstrate the kinds of new opportunities that are emerging for farmers' cooperatives to respond to the growing demand for high-quality and organic food (see Chapter 3), we selected cases of cooperatives adopting "alternative" farming strategies. We recognize, however, that most cooperatives in China are still oriented to conventional agricultural production. Moreover, we chose to focus on successful cases to better understand the contributions of cooperatives to rural development. Cooperatives initiated by agribusinesses are not included in this study because agroenterprises are mainly driven by profit maximization rather than a rural development goal. Clegg (2006) found that the agroindustrial-oriented model in China leads to the monopolization of benefits by wealthy farmer-investors and outside shareholders at the expense of small-scale farmers. Moreover, this model does not address the disadvantaged position of small-scale farmers in decision-making and in the distribution of earnings (see Yan and Chen 2013 regarding the debates over rural cooperatives in China).

We conducted over 20 face-to-face semi-structured interviews for this part of our overall study. Interviewees were selected using purposive sampling and included the cooperative initiators, cooperative leaders, cooperative members, and organic certification agencies, as well as other key actors such as representatives from local institutions and government agencies. At least four interviews focused specifically on each of the FPC cases, including one with each cooperative leader. In addition to this interview data, we also reviewed secondary sources in this study, including government reports, project reports of organic agricultural development in less developed regions, and cooperative documents of the Tonglu and Daizhuang FPCs.

Farmers' cooperatives and sustainable rural development

Three distinct agrarian production paradigms can be distinguished to highlight the transformation toward sustainability of rural development: the "agroindustry paradigm," the "post-productivist paradigm," and the "sustainable rural development paradigm" (Marsden 2003). Based largely on Europe's experience, these three paradigms differ in internal logic, ideology, scientific rationality, and regulatory arrangement (Marsden, Banks, and Bristow 2002). The agroindustrial paradigm, following the logic of neoclassical economics, promotes specialization and economies of scale. The post-productivist paradigm is based on the belief that the agricultural sector (in developed economies) is being marginalized through a move away from food production and toward the "consumption" of the countryside (Marsden et al. 1993). Marsden (2003) argues that both of these two development paradigms are unsustainable.

The new sustainable rural development paradigm, in contrast, redefines our relationship with nature by highlighting the multifunctionality of agriculture and works toward an alternative food supply chain to counter the scale and price rationalities of large-scale agribusiness (Marsden, Banks, and Bristow 2002). The rural development paradigm explores opportunities in the agricultural sector related to resource use, livelihood strategies, and institutional arrangements. This paradigm reasserts land-based agricultural production as a central dimension in achieving rural sustainability and highlights the crucial roles of farmers and farmers' cooperatives in revitalizing the rural economy (van der Ploeg et al. 2000; Marsden, Banks, and Bristow 2002). It emphasizes the ability and skills of farmers and FPCs to generate different economic values from the same ecological resource through co-production, cooperation and co-evolution of the resource base (Marsden 2009). The multifunctional role of agriculture in meeting new social and environmental demand is underlined in this paradigm (Renting et al. 2009; van der Ploeg et al. 2009).

Part of the debates over the rural development paradigm concerns the role and categorization of rural development activities. Although this paradigm has been widely used, there is no comprehensive and agreed upon definition of it (van der Ploeg 2000). To identify an activity as a "rural development activity," Marsden, Banks, and Bristow (2002) postulate that the aggregated effect of this activity must meet the following three conditions: (1) It is a response to the cost-price squeeze on agriculture and adds income and/or employment opportunities to the agricultural sector; (2) it corresponds to the needs and expectations of the population and expresses new relationships between the agricultural sector and society; and (3) it implies a redefinition, recombination, and/or reorganization of rural resources and develops new businesses and/or opportunities within rural society. The diversified activities can take place on-farm and/or within the local economy, either within the scope of agriculture or outside of it (van der Ploeg et al. 2002). Beyond the production of raw materials, alternative activities include landscape management, agritourism, innovative forms of cost reduction, production of high-quality and region-specific products, direct marketing, and new activities such as care activities for the disabled (Darnhofer 2005: 309).

Diverse rural development activities have been clustered into a typology of "deepening," "broadening," and "regrounding" alternative farming strategies by van der Ploeg et al. (2002) (see Figure 4.1). A *deepening* strategy refers to activities that add value to products by means of processing or by focusing on "quality" production (such as organic) or shortening the food supply chain. A *broadening* strategy refers to activities that diversify non-agricultural activities based on rural resources, such as agritourism and landscape conservation. A *regrounding* strategy refers to activities that reorganize farm resources mainly through reallocating family labour, reutilizing farm resources, or adopting various forms of local and regional cooperation and/or collaboration to achieve cost reductions.

With a few exceptions (van der Ploeg, Ye, and Schneider 2012), this framework has been applied to date mainly within EU rural development contexts (see for example Ortiz-Miranda, Moreno-Pérez and Moragues-Faus 2010). We adopt this typology for our study with two slight modifications: (1) We consider green, hazard-free,[1] and organic agriculture as ecological agriculture under the category of deepening strategy in this chapter; and (2) we do not consider off-farm income to be a regrounding strategy for FPCs. Part-time farming is a common phenomenon in rural China, so it should not be viewed as an alternative farming activity. Moreover, the effects of part-time farming on rural development in China are contradictory, as we explain later in this chapter.

The farmers' cooperative model provides an important lens to analyze rural development at the regional level, although to date this model has received little attention in rural development research (Ortiz-Miranda, Moreno-Pérez, and Moragues-Faus 2010; Knickel and Renting 2000). Following the deepening–broadening–regrounding typology, we examine the potential contributions of FPCs to agricultural multifunctionality and rural development in China. The deepening–broadening–regrounding typology provides an analytical framework for describing and assessing agricultural multifunctionality and rural sustainability. Beyond producing food and

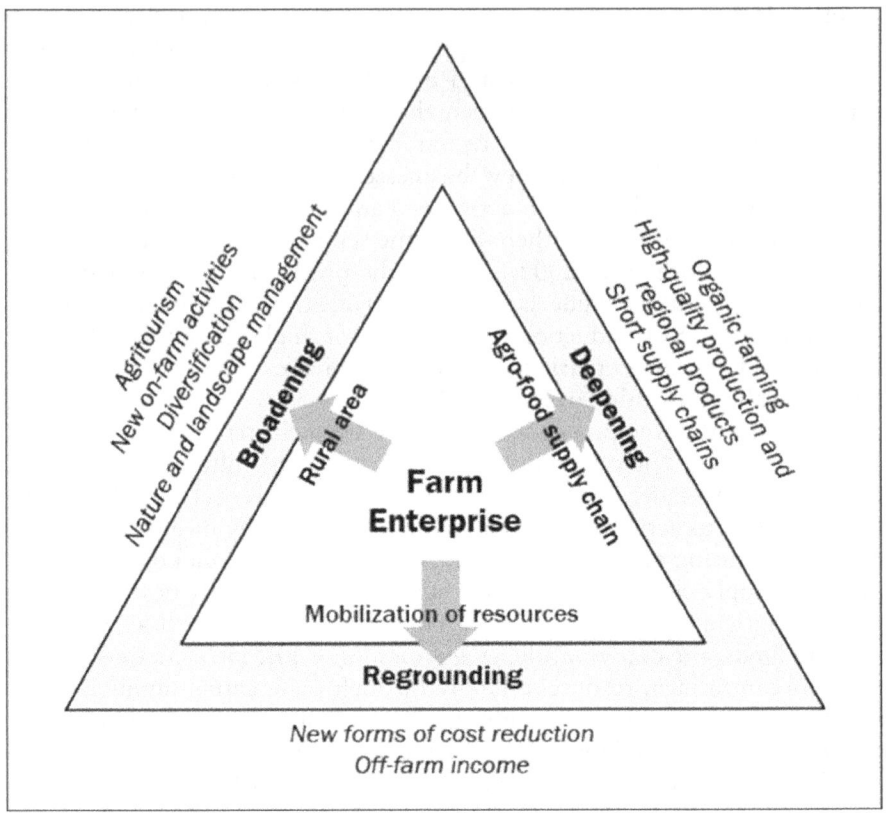

Figure 4.1 Boundary shifts: the "deepening–broadening–regrounding" typology.

Source: van der Ploeg, Long and Banks (2002), as cited in van der Ploeg et al. (2012: 134).

fibre, and providing employment and income, agriculture is considered to be one of the most common multifunctional activities, as it also produces other commodities (such as agritourism and other services) and non-commodity outputs (such as landscape management, soil conservation, and biodiversity) (Durand and van Huylenbroeck 2003; Renting et al. 2009).

The evolution of farmers' cooperatives in China

Cooperatives around the world have been a central institution in social development, poverty alleviation, employment creation, and participatory development (United Nations 2001). The cooperative model is defined by the International Co-operative Alliance (ICA) as "an autonomous association of persons united voluntarily to meet their common economic, social, and cultural needs and aspirations through jointly-owned and democratically-controlled enterprise" (n.d. para. 1). Cooperatives can deliver pro-poor growth in a manner that is owned and controlled by poor and small-scale farmers themselves (Clegg 2006). Nevertheless, farmers' cooperatives, particularly in the Global South, face many challenges owing to the lack of capital and business management capacity (Birchall 2004).

Looking back over the evolution of farmers' cooperatives since the establishment of the People's Republic of China in 1949, we can distinguish three phases: from 1949 to the early 1980s, the early 1980s to 2007, and 2007 to the present. Since 1949, agrarian institutions have changed from agricultural "collectives" or people's communes in the Mao era[2] to family farming and then to FPCs (Jia et al. 2010). The unsuccessful experience of agricultural collectives during the Maoist period became an obstacle to developing farmers' cooperatives in the following decades (Hayward 2017). The level of trust among people—an important basis for cooperation—was eroded in many systems of collectivization owing to centralized decision-making that left little or no room for civil society initiatives and social organizations (Paldam and Svendsen 2001). Subsequent challenges have been reported in some post-socialist countries with (re-)establishing farmers' cooperatives (see Paldam and Svendsen 2001; Tisenkopfs et al. 2010). Agricultural collectives in China stagnated between the 1960s and early 1980s. In the late 1980s, with the shift from central planning to a market economy orientation, agricultural cooperatives began to emerge, particularly in the fruit and vegetable sectors (Garnevska, Liu and Shadbolt 2011; Zhang 1999). These cooperatives mainly involved pre- and post-farm production activities in relation to purchasing farm inputs, processing, and marketing (Clegg 2006).

In Japan, South Korea, and Taiwan, experiences of the "East Asia development model" indicate that rural development often garners more attention when the industrialization and urbanization of a country reach a certain phase. To build a stronger rural community and improve the living conditions of rural households, community-based rural development initiatives, especially farmers' cooperatives, have been promoted in these countries through policy support (Choi et al. 2007; Long et al. 2010; Suh 2015; Hayward 2017).

China's fast-growing economy, and stronger international standing, meant that by the 2000s China was in a position to broaden its development strategy and provide more support to agricultural and rural development (Long et al. 2010). It was against this backdrop that the first national Farmers' Professional Cooperative Law was enacted in 2007 to formalize and standardize FPCs in China. The law stipulates that FPCs must be voluntarily and democratically organized and remain independent in operation. Having FPCs controlled democratically by farmers sets them apart from the previous agricultural collectives of the socialist era, in which the supplying of farming inputs and producing and selling activities were all centrally planned by government (Hu et al. 2004).

This stable legal environment together with various supportive government policies has created a favourable political and economic environment for developing FPCs in China. As a result, the number of FPCs has been increasing rapidly since 2007 (see Table 4.1). However, as noted earlier, most FPCs are criticized for being "fake" cooperatives that are controlled by a small group of members with the purpose of gaining state subsidies and fail to empower small producers in practice (Yan and Chen 2013). The "fake" cooperatives, mainly those initiated by agroenterprises, are different from the types that we examine in this study. Meanwhile, cooperation among FPCs across multiple townships is also developing in China; this increases their market power and provides more services for farm members (Garnevska, Liu, and Shadbolt 2011). The main activities defined in the Cooperative Law include purchasing agricultural inputs, marketing, processing, transportation, storage, and providing agricultural technology and information.

Learning from the experience of "comprehensive cooperation" in Japan, South Korea, and Taiwan, many Chinese rural development advocates and intellectuals (e.g. Wen Tiejun and Li Changping) also highlight the values of FPCs in empowering rural areas and small producers rather than focusing only on commodity production (Yan and Chen 2013; Song et al. 2014). In situations where farmers are poorly educated, lack cooperative management experience, and have limited access to legal advice, intellectuals who advocate for rural development have called on the Chinese government (at both national and local levels) to play a stronger role in promoting and organizing FPCs (Yang and Wen 2011; Yang et al. 2017). With the strong support that the Chinese government has been giving to large-scale agribusiness enterprises, in particular "dragon-head enterprises," since the mid-1990s, the capacity of cooperatives has suffered

Table 4.1 Growth in the number of farmers' cooperatives in China, 2007–2018

	# of registered FPCs	# of registered members	Registered capital (millions of US$[a])
2007	26,400	350,000	5,074
2008	110,900	1,417,100	14,329
2009	246,400	3,917,400	40,070
2010	379,100	7,155,700	74,002
2012	521,700	11,964,300	117,950
2014	1,288,800	92,270,000	402,989
2018	2,044,000	117,590,000	690,693

Sources: State Administration for Market Regulation (for 2014 and 2018) and Fleischer, 2012: 24, using data from Ministry of Industry and Commence and the General Station of Administration on Rural Cooperative Economy, Ministry of Agriculture.

Note
a One US$ was valued at 6.14 Chinese yuan as of August 2014.

(Wen and Dong 2010). Yang and Wen (2011) call for stronger government support for developing cooperatives that "ensure fairness and protect the disadvantaged" and "represent integrative and long-term social interests" (p. 45).

There is a rather uneven geography of development of farmers' cooperatives across China. Cooperatives developed first and most rapidly in eastern China, where the economy and markets were more developed and agriculture is more industrialized, whereas cooperatives in less-industrialized western China has been somewhat slower (Liang and Hendrikse 2013). Zhejiang is a pioneering province in eastern China where the first modern farmers' cooperatives in China were established. Zhejiang takes the lead in the development of farmers' cooperatives in China, in terms of both the total number of FPCs and their economic performance (Liang and Hendrikse 2013; Sultan and Larsén 2011). It was also a leading province in enacting the provincial Cooperative Law and regulations in 2005, providing the basis for the national law[3] promulgated on July 1, 2007.

Cooperative types

Cooperative profiles

Daizhuang Organic FPC: Daizhuang village, south of Jurong city in Jiangsu province, is situated on hilly land with 1,040 hectares or 2,570 acres (approximately 666.7 ha or 1,648 acres of farmland, 60 per cent of which is hilly-slope land) and a population of 2,900 (around 866 households). At the time of establishing the cooperative, it was the poorest village within Zhenjiang city, despite boasting rich natural resources. Before the establishment of the Daizhuang Organic FPC, conventional crops, including wheat and rice, were produced. After a comprehensive study, a senior researcher, Mr. Zhao at the Institute of Zhenjiang Agricultural Technology and Science (IZATS), facilitated the establishment of this cooperative in 2006. Since that time, Mr. Zhao has continued to serve as an on-site technical consultant, and the village secretary has served as the cooperative leader,[4] attending to the daily management of activities in the cooperative. Daizhuang Organic FPC was the first organic farmers' cooperative in Jiangsu province. Its main products are organic rice and strawberries. Products are sold through various channels, including direct sale to companies[5] (60 per cent of sales), to individuals (20 to 30 per cent), through its own specialty stores locally, and via agencies in large cities. Home delivery was offered in 2007 and 2008 but was discontinued owing to the high cost. Given the small volume of production, this cooperative faces challenges in supplying a large food retailer.[6]

Tonglu Peach FPC: Yangsanfan village, in the northern part of Tonglu county, in Zhejiang province, is situated in a mountainous area with 519 hectares or 1,282 acres (approximately 74 ha or 183 acres of farmland and 155 ha or 383 acres of forest land) and a population of 861 (285 households). Peaches have been grown in this area for approximately 170 years. Compared to other areas in China, rural communities in Zhejiang province are wealthier and farmers have greater entrepreneurial skills. The per capita income in this village was around US$2,000 in 2008. With the support of local government agencies, the Tonglu Peach FPC was initiated in 2004 by a few local "large-scale" peach farmers[7] and was the first farmers' cooperative in Tonglu County. Mr. Wang, one local large-scale peach farmer, has acted as the elected cooperative leader since it was founded. He is high school–educated and is active in marketing and establishing social networks. Peaches and cherries are the main products of this cooperative. Peaches are sorted into two grades: first-class peaches are gift packaged and are procured by companies and government agencies as gifts for employees[8] or are sold at specialty fruit markets in large cities; second-class peaches are sold at wholesale markets. Agritourism is also a channel for this cooperative to sell its products.

Yuexi Organic Kiwifruit FPC: Yufan village in Yuexi county, Anhui province, is situated in a cool mountainous area with 950 hectares or 2,348 acres (approximately 95 ha or 235 acres of farmland, 68 per cent of which is paddy field and the rest is dry land, and 850 ha or 2,100 acres of forest land) and a population of 1,005 (257 households). It is the poorest village in the area. The Yuexi Organic Kiwifruit FPC[9] was established in 1999 in Yufan village with the support of a Deutsche Gesellschaft für Technische Zusammenarbeit (GTZ) project,[10] the Organic Food Development Center (OFDC),[11] and the local government. The cooperative produced organic kiwifruit and water bamboo. Mr. Chu, a former village officer, has served as the elected cooperative leader because he knows the local situation well and is willing to devote himself to local development. Following the end of GTZ project support in 2003, the organic kiwifruit FPC was divided into two groups in 2006: the kiwifruit FPC and the water bamboo FPC. The latter has been growing rapidly. The withdrawal of the GTZ project posed a big challenge to the kiwifruit FPC to continue organic farming because of the high certification costs, a shortage of funding, and limited access to value-added markets to garner a sufficient price premium. As a result, the FPC discontinued organic kiwifruit farming. Organic kiwifruits had been exported with the assistance of the GTZ project, while non-organic kiwifruits have been sold domestically through various channels since the project support ended. Water bamboo is delivered to large cities (e.g. Shanghai, Nanjing, Hefei) and sold at wholesale markets. More recently, the retirement of the kiwifruit FPC leader also created difficulties as members lacked confidence in the new leader. Key characteristics of the three FPCs are summarized in Table 4.2.

Table 4.2 Key characteristics of three professional farmers' cooperatives in China[a]

	Daizhuang FPC (Jiangsu province)	Tonglu Peach FPC (Zhejiang province)	Yuexi FPC (Anhui province)
Locations	Close to large cities (Nanjing and Shanghai)	Close to large cities (Hangzhou and Shanghai)	Far from large cities
Initiators	Several local farmers with large-scale farmland	A researcher	The GTZ Project and the Organic Food Development Center
Leaders	A large-scale farmer	Local government official	Former local government official
Year established	2006	2004	Founded in 1999 and registered in 2001
Number of members	612 households (70% of village households) in 2012; 3 households in 2006	173 households (60% of total) in 2011; 116 households in 2004	No updated data (138 households in 2002; 43 households in 2001)
Technical innovation	Introduced a new rice variety from Japan	Applied new technology to stagger the harvest time	Promoted suitable crops for local natural resources
Main products	Organic rice and strawberries	Peaches (green and hazard-free certified) and cherries	Kiwifruit and water bamboo (hazard-free certified)[b]
Target markets	Domestic; various channels	Domestic; gift packages and wholesale markets	International (only for organic kiwifruit) and domestic; wholesale

Notes

a Most data listed in the table were supplied through interviews; additional information came from the report of the Yuexi Organic development Project 2002 (Bao, 2002) and cooperative documents of the Tonglu and Daizhuang FPCs.

b Most of the arable land in Yuexi county is cold, waterlogged paddy field, which is not suitable for growing regular crops (i.e. rice) and has low yields, but it is ideal for growing water bamboo. The Yuexi FPC took advantage of the local natural conditions and encouraged farmers to grow water bamboo.

Alternative strategies and the new rural development paradigm

FPCs provide a variety of services for their members—services that support on-farm activities (such as providing technical assistance and purchasing inputs together) and/or facilitate marketing their produce (such as sorting, grading, marketing, and processing). Activities and strategies adopted by the three FPCs can be grouped into the three categories of deepening, broadening, and regrounding (Table 4.3).

Deepening strategy: Following Renting, Marsden, and Banks (2003), these initiatives can be considered to be new configurations of AFNs. All three FPCs have undertaken such initiatives to increase the value of their products. Three main types of deepening strategies were pursued. First, product branding was developed by all three FPCs with the goal of improving the reputation and market competitiveness of their products. Second, ecological and local characteristics of products (certified organic, green, hazard-free and geographical identification) were simultaneously highlighted in all three FPCs. These formalized standards and labels show the attributes of product quality and can help diversify marketing channels (Renting, Marsden, and Banks 2003). Although organic certification was not continued in the Yuexi FPC after the GTZ project withdrew in 2003, all water bamboo produced in Yuexi county is hazard-free and geographical identification–certified. The Tonglu cooperative received hazard-free certification for 200 ha (949 acres) in 2005 and green certification for 67 ha (166 acres) in 2006. Peaches were sorted into two grades: first-class peaches for gift packages and second-class peaches for wholesale markets. The third type of deepening strategy, employed by the Daizhuang and Tonglu FPCs, was to use used direct marketing strategies to sell most of their produce. The Yuexi FPC did not, owing to its long distance from customers.

Broadening strategy: At the time we conducted interviews, the Tonglu FPC was the only one among these three cases that developed a broadening strategy, although the leader of the Daizhuang FPC expressed strong interest in promoting agritourism. With the support of the Tonglu municipal government, the Tonglu FPC collaborated with several other FPCs in the same area to host visitors during the period of the Flower Festival (lasting for four months from late March to mid-July). During the festival period, they organized many activities, including cultural performances, demonstrations of local agricultural products, tastings, sales, signing sales contracts, picking local fruits, and homestays with rural households. Agritourism (also called "agritainment"—experiencing life in a rural area) has become a popular form of rural tourism for many urbanites in China (Marsden, Yu, and Flynn 2011). The leader of the Daizhuang FPC also viewed agritourism as a potential channel to sell its produce by hosting harvest festivals or other activities; it plans to develop agritourism in the near future. Agritourism was not mentioned in the Yuexi FPC, likely owing to its distance from urban areas.

Table 4.3 Typology of strategies pursued by the three cooperatives

	Daizhuang FPC	*Tonglu Peach FPC*	*Yuexi FPC*
Deepening strategy	Product branding; food processing; organic certification; direct marketing	Product branding; sorting and packaging; green and hazard-free certification; direct marketing	Product branding; organic certification; hazard-free certification; geographical identification (GI) of raw materials (water bamboo)
Broadening strategy	Plan to develop agritourism	Flower Festival and agritourism	None
Regrounding strategy	Unified farming management	Unified farming management; collaborating with other FPCs in the same region to develop agritourism	Unified farming management; collaborating with other FPCs in the same region to transport products to larger cities

Regrounding strategy: All three FPCs have developed and implemented a regrounding strategy in the form of unified farming management, which can reduce production and transaction costs on member farms by taking advantage of economies of scale. The FPCs made unified plans for farming activities (i.e. what, when, and how it is produced) to enable an adaptive response to increasingly differentiated market demand (such as quality requirements, seasons, product presentation). They also provided various services to their members, such as technical assistance and training; supplying ecological fertilizers and pesticides; supplying seeds and seedlings; and product processing, packaging, and marketing. Collaborations among FPCs in the same region were adopted by the Yuexi and Tonglu FPCs to reduce the costs for transportation and for hosting events, respectively, even though these collaborations were informal and very loose. The leader of the Daizhuang FPC planned to combine crop cultivation and breeding (geese, in this case) to offset the low productivity of organic rice farming and to increase farmers' income.

The broadening activities are far less developed across all three FPCs compared to the deepening activities. The only exception is the Tonglu FPC's agritourism endeavours. The adoption of non-agricultural activities is more challenging for FPCs. The underdevelopment of the broadening activities can be explained by the following reasons. First, some initial conditions are required to develop these kinds of activities. For example, an initial but significant investment is needed for developing and organizing agritourism in making rural areas attractive, such as providing accommodation facilities, arranging activities, and offering suitable opportunities for spending (Gannon 1994). Considering the significance of the investment and the uncertainty of economic returns, an FPC is often unable or reluctant to invest in these facilities. Second, government needs to play an important role in funding and facilitating agritourism at the initial stage (Fleischer and Felsenstein 2000, Iorio and Corsale 2010). This poses challenges for developing agritourism in poor areas (for instance, Anhui province in our case study), where the local government has a more limited budget. The third reason for the underdevelopment of the broadening activities is that, besides economic constraints, developing non-agricultural activities and in particular agritourism often requires new skills such as guest services, marketing, and advertising (Sharpley 2002), which are unfamiliar to farmers. In addition, collaboration among FPCs in the same region is important in developing agritourism, as we saw in the Tonglu FPC case (see also van der Ploeg, Ye, and Schneider 2012).

Feeding into rural development processes

This chapter has examined a series of diversified strategies adopted by three farmers' professional cooperatives engaging in ecological agricultural production in three provinces of China. These activities have a range of

different expressions, including capturing greater value-added in production via certification, branding, processing, sorting, and packaging (found
in all three FPCs); shortening supply chains (e.g. providing home delivery
and operating local specialty stores in the Daizhuang FPC); and expanding
to other on-farm activities (e.g. agritourism in the Tonglu FPC). Rural
systems with strong multifunctionality can offer diverse opportunities for
residents in terms of earning non-agricultural income (e.g. agritourism),
maintaining high environmental quality, and increasing stakeholder
involvement and rural democracy (Wilson 2010). We categorized the
diversified rural development activities into three alternative strategies:
deepening, broadening, and regrounding. We assessed the economic,
social, and environmental impacts of farmers' cooperatives associated with
adopting these activities and strategies. This provided a sense of their contributions to agricultural multifunctionality and rural development.

The economic contribution of FPCs to rural development is significant
in all three cases. Members in all three FPCs have reported a significant
increase of their household income from agricultural production. For
example, the average household income of members of the Daizhuang FPC
increased by approximately RMB5,000 (US$310) in 2010. By taking
advantage of economies of scale FPCs help overcome the limitations of
small-scale farming in terms of supplying input, marketing outputs, reducing transaction costs, enhancing the safety and quality of agricultural production, increasing market competitiveness, and expanding new markets
or value chains. The "deepening" activities enhance the economic empowerment of small-scale farmers by linking them to value-added markets (e.g.
ecological and organic products, branding, processing, sorting, and packaging) (see also Qiao et al. 2018). Beyond producing food, the Tonglu FPC
also adopted a broadening strategy (i.e. agritourism) to help advertise the
cooperative and increase the reputation of its products. Through united
management and collective decision-making, the "regrounding" activities
provide economic contributions to farm members by reducing production
and transaction costs and responding more effectively to market demand.
These diversified activities contribute significantly to improving household
incomes and living conditions of cooperative members, which are also the
goals of current agricultural policies.

All three FPCs have experienced substantial growth in cooperative membership since their establishment. As the leader of the Tonglu FPC explained,

> Since our cooperative was founded, many strategies have been
> adopted, such as branding, certification, sorting and packaging, direct
> marketing, etc. These strategies have helped increase the prices of our
> products. Our members now receive higher economic returns from
> farming. So farmers in our village and those in surrounding villages all
> want to join in our cooperative. But our cooperative only accepts new
> members who meet our stringent selection criteria, like willingness to

follow the cooperative rules and our production standards, self-discipline, etc.

(Interview with the manager of the Tonglu FPC in Zhejiang, China, March 24, 2011)

Economic benefits of the cooperatives are not distributed equally among members. This can be attributed to the fact that farmers differ in their financial assets, skills, and social networks. In addition to selling agricultural products to the FPC, some core members also invest capital in the FPC that gets used for purchasing inputs, processing and sorting machines, and cold storage facilities. They have both user shares and investor shares[12] in the FPC. Therefore, these core members often hold more shares and correspondingly benefit more from the FPC, whereas common members only benefit by selling their products to the FPC (see also Liang and Hendrikse 2013).

Conclusions

Our findings point to the important social contributions that farmers' cooperatives in China have made to rural development. The social contributions revealed in our case studies can be categorized into four aspects: social integration, local and regional embeddedness, adoption of food quality standards and food safety, and rural democracy and governance. We will discuss each of these in turn. First, in terms of social integration, on the one hand, the farmers' cooperative model provides a platform for farm members to exchange experiences and gain new knowledge, which further reinforces the ties and enhances social integration among members. On the other hand, through collaborating with other cooperatives, universities, and research institutes, farmers' cooperatives have enhanced their capacity to network with other actors. However, in our case studies we found that the integration among cooperatives was still very loose, partially because the newly enacted Cooperative Law does not define a cooperative federation (i.e. a supra-cooperative network). This omission could pose significant constraints for cooperatives to grow and gain strength in the global market (Fleischer 2012).

In terms of local and regional embeddedness—the second aspect of social contributions to rural development—direct marketing strategies adopted by the Tonglu and Daizhuang FPCs helped reconnect producers and consumers and renegotiate the trust relationship between them. This further contributed to high levels of social embeddedness and relations of regard (Hinrichs 2000; Milestad et al. 2010). The degree of local and regional embeddedness of the food supply chain is an important indicator of rural development (Knickel 2001) and a strongly multifunctional agriculture regime (Clark 2003).

Third, each cooperative in our case study adopted certain types of food quality production standards and registered a brand for their products, which helped to ensure a safer food supply in China (see also Jin and Zhou 2011). In addition, as the main actors in FPCs, farmers gained experience in cooperation and democratic governance by electing cooperative leaders and participating in decision-making (although this was limited to the production domain for common members in our case studies).

Environmental contributions of farmers' cooperatives to rural development were also evident in all three FPCs. All three engaged in ecological agriculture (green, hazard-free food and organic agriculture in our cases), which helps to build soil fertility and minimize environmental externalities. Localized food supply chains established by the Daizhuang and Tonglu FPCs reduce the distance that food travels from the site of production to consumption, thereby reducing the need for long-distance food transport and its associated energy emissions (Goodman 2004). Agritourism can help improve the awareness of environmental problems among both farmers and urban visitors (Brodt et al. 2006).

Although FPCs have developed rapidly in China over the past decade, progress has not been uniform across the country owing to differences in farmers' education levels and varying economic and social situations among different regions of the country (Garnevska, Liu, and Shadbolt 2011), as well as varying levels of government support and of trust among farmers. FPCs face many challenges for developing further.

Several conclusions can be drawn from the analysis of these three cases. First, in adopting the "deepening–broadening–regrounding" typology (van der Ploeg et al. 2002) for our analysis, we found that the deepening and regrounding strategies were more commonly applied by all three FPCs than the broadening strategy. Broadening activities, such as agritourism, are more challenging for China's FPCs because of the requirements for capital investment and new management and marketing skills. Second, our case studies demonstrate the potential of FPCs to make significant economic, social, and environmental contributions to rural development. However, our interviews suggest that economic gains are not shared equally among members in the cooperative. Common members only benefit by selling their products to the cooperative, whereas core members can benefit by both selling their products to and investing capital in the cooperative.

Notes

1 Given that GMO and certain types of pesticides and fertilizers are allowed in production, green and hazard-free production practices would not be considered ecological agriculture in a European or North American context. We categorize green and hazard-free production practices under "ecological agriculture" sector here because they have a tendency toward reducing ecological impact by limiting (somewhat) the amounts and types of agrochemicals used, as compared to conventional farming practices in China.

2 We use the phrase "collective" here to refer to the type of collective action with the purpose to overcome barriers faced by individual farms. Although in the

Chinese literature "collective" is sometimes translated into English as "cooperative," we recognize that "collective farms" in the Mao era would not be considered cooperatives today. The "collectives" in the Mao era did not meet the criteria of cooperatives, such as being voluntary to join or withdraw.

3 In China, provinces or municipalities are allowed and selected (in some cases) to experiment with new projects or strategies in a given area, and then the state learns from this and the experience shapes the national law. This differs from the procedure in many other countries, where a law is enacted and then people follow it in a much more linear system than in China (also see van der Ploeg, Ye, and Schneider 2012).

4 Given that many young people have migrated to urban areas for better job opportunities, secretaries are often the ones with a better education and stronger ability than others in rural areas.

5 In China, it is common for an employer to purchase gifts for employees or clients on special occasions throughout the year.

6 The large retailer refers in particular to Beijing Organic and Beyond Corporation (OABC), which is one of the largest companies engaging in the cultivation, production, distribution, and home delivery of organic food in China. Although this company has its own production bases, it also buys organic products from enterprises or cooperatives. FPCs also face great difficulties in selling their products through supermarkets, the major food outlets in most cities, partially due to the small volume of their production and the high standards that must be met. It is also costly to sell products through supermarkets, including paying stocking, sales, and promotional fees, and giving 20 per cent of the profit to the store (Lagos et al. 2010). Therefore, many FPCs choose to sell their products at wholesale markets or via direct marketing channels (as we illustrated in the three cases discussed here).

7 We recognize that there are significant differences in definitions and in the understanding of what constitutes a small- versus large-scale farm in China and the West. In this study, small-scale farming refers to Chinese small-scale farms with an average size of less then 0.5 hectares or 1.2 acres per household, whereas large-scale farming refers to farm sizes over 1.3 hectares or 3.2 acres. During our interviews from 2010 to 2011, farms with sizes over 20 mu (or 1.3 hectares or 3.2 acres) were referred to by several cooperative leaders as large-scale farms. Some of these farms lease land from their relatives or neighbours who choose to work in non-agricultural sectors in cities; others lease undeveloped village land from rural collectives. The latter often have comparatively larger scales (e.g. over 50 mu or 3.3 hectares or 8.2 acres), as we have seen in the Tonglu case.

8 This cooperative, collaborating with several other cooperatives that produce different crop varieties in the same area, runs its own specialty stores and attracts local consumers.

9 This cooperative is supported by Yufan Kiwifruit Research Institute, which was founded by several local farmers in response to serious plant diseases and insect pests suffered by kiwifruit farmers in the village from 1991 to 1993. With the technical support from the institute, kiwifruit production grew rapidly in the following ten years, and this area became "the first township of kiwifruit production in East China" with over 290 ha (717 acres) under kiwifruit cultivation.

10 The Sino-German GTZ project (1998–2003), named "Development of Organic Agriculture in Poverty Areas in China," was initiated to offer an advisory service and information system in China for organic agricultural development.

11 The Organic Food Development Center (OFDC), founded in 1994 in Nanjing by the former Chinese State Environmental Protection Agency, is the first specialized organization engaged in research, certification, training, and promotion of

organic agriculture in China. It is also one of the largest certification bodies in China.

12 According to the FPC law, no single member can hold more than 20 per cent of the total investor share in the cooperative.

References

Birchall, J 2004, *Cooperatives and the Millennium Development Goals*. ILO, Geneva.

Brodt, S, Feenstra, G, Kozloff, R, Klonsky, K and Tourte, L 2006, "Farmer-community connections and the future of ecological agriculture in California," *Agriculture and Human Values*, vol. 23, no. 1, pp. 75–88.

Choi, J, Kim, J, Kim, J, and Kim, J 2007, *Improved Community Driven Development (CDD) Strategy: The Case of New Village Movement in Korea*.

Clark, J 2003, "Regional innovation systems and economic development: The promotion of multifunctional agriculture in the English East Midlands," PhD thesis, Department of Geography, University College London, UK.

Clegg, J 2006, "Rural cooperative in China: Policy and practice," *Journal of Small Business and Enterprise Development*, vol. 13, no. 2, pp. 219–234.

Darnhofer, I 2005, "Organic farming and rural development: Some evidence from Austria," *Sociologia Ruralis*, vol. 45, no. 4, pp. 308–323.

Durand, G and van Huylenbroeck, G 2003, "Multifunctionality and rural development: A general framework." In: van Huylenbroeck, G and Durand, G (eds.), *Multifunctional Agriculture: A New Paradigm for European Agriculture and Rural Development*, Ashgate, Aldershot, UK, pp. 1–18.

Fleischer, G 2012, "Between organic development and state control," *Rural Focus*, vol. 21, pp. 23–25.

Fleischer, A and Felsenstein, D 2000, "Support for small-scale rural tourism: Does it make a difference?" *Annals of Tourism Research*, vol. 27, no. 4, pp. 1007–1024.

Gannon, A 1994, "Rural tourism as a factor in rural community economic development for economies in transition," *Journal of Sustainable Tourism*, vol. 2, no. 1 + 2, pp. 51–60.

Garnevska, E, Liu, G, and Shadbolt, N 2011, "Factors for successful development of farmer cooperatives in Northwest China," *International Food and Agribusiness Management Review*, vol. 14, no. 4, pp. 69–84. Accessed at http://agecon-search.umn.edu/bitstream/117603/2/20110028_Formatted.pdf on May 9, 2018.

Goodman, D 2004, "Rural Europe redux? Reflections on alternative agrofood networks and paradigm change," *Sociologia Ruralis*, vol. 44, no. 1, pp. 3–16.

Gürel, B 2014, "Changing relations of production in Chinese agriculture from decollectivization to capitalism," *McGill Sociological Review*, vol. 4, pp. 67–92.

Hale, M 2013, "Reconstructing the rural: Peasant organizations in a Chinese movement for alternative development," PhD thesis, University of Washington, Seattle, WA.

Hayward, J 2017, "Beyond the ownership question: Who will till the land? The new debate on China's agricultural production," *Critical Asian Studies*, vol. 49, no. 4, pp. 523–545.

Hinrichs, C 2000, "Embeddedness and local food systems: Notes on two types of direct agricultural market," *Journal of Rural Studies*, vol. 16, no. 3, pp. 295–303.

Hu, D, Reardon, T, Rozelle, S, Timmer, P, and Wang, H 2004, "The emergence of supermarkets with Chinese characteristics: Challenges and opportunities for China's agricultural development," *Development Policy Review*, vol. 22, no. 5, pp. 557–586.

Huang, P 2011, "China's new-age small farms and their vertical integration: Agribusiness or co-ops?," *Modern China*, vol. 37, no. 2, pp. 107–134.

Iorio, M and Corsale, A 2010, "Rural tourism and livelihood strategies in Romania," *Journal of Rural Studies*, vol. 26, no. 2, pp. 152–162.

Jia, X, Hu, Y, Hendrikse, G, and Huang, J 2010, *Centralized versus Individual: Governance of Farmer Professional Cooperatives in China*. Paper presented for presentation at the IAMO Forum: Institutions in Transition: Challenges for New Modes of Governance, Halle (Saale), Germany, June 16–18, 2010.

Jin, S and Zhou, J 2011, "Adoption of food safety and quality standards by China's agricultural cooperatives," *Food Control*, vol. 22, no. 2, pp. 204–208.

Knickel, K 2001, "The marketing of Rhongold milk. An example of the reconfiguration of natural relations with agricultural production and consumption," *Journal of Environmental Policy and Planning*, vol. 3, no. 2, pp. 123–136.

Knickel, K and Renting, H 2000, "Methodological and conceptual issues in the study of multi-functionality and rural development," *Sociologia Ruralis*, vol. 40, no. 4, pp. 512–528.

Kruijssen, F, Keizer, M, and Giuliani, A, 2009, "Collective action for small-scale producers of agricultural biodiversity products," *Food Policy*, vol. 34, no. 1, pp. 46-52.

Lagos, JE, Scott, RR, Rasmussen, K, Bugang, W, and Chen, U 2010, *Organic Report, China—People's Republic of* (GAIN Report No. 10046). Accessed at http://gain.fas.usda.gov.

Lammer, C 2012, "Imagined cooperatives: An ethnography of cooperation and conflict in new rural reconstruction projects in a Chinese village," thesis, Department of Social and Cultural Anthropology, University of Vienna.

Liang, Q and Hendrikse, G 2013, "Core and common members in the genesis of farmer cooperatives in China," *Managerial and Decision Economics*, vol. 34, no. 3–5, pp. 244–257.

Long, HL, Liu, YS, Li, XB, and Chen, YF 2010, "Building new countryside in China: A geographical perspective," *Land Use Policy*, vol. 27, no. 2, pp. 457–470.

Marsden, T 2003, *The Condition of Rural Sustainability*. Royal Van Gorcum, Assen, the Netherlands.

Marsden, T 2009, "Mobilities, vulnerabilities and sustainabilities: Exploring pathways from denial to sustainable rural development," *Sociologia Ruralis*, vol. 49, no. 2, pp. 113–131.

Marsden, T, Banks, J and Bristow, G 2002, "The social management of rural nature: Understanding agrarian based rural development," *Environment and Planning A*, vol. 34, no. 5, pp. 809–825.

Marsden, T, Murdoch, JL, Lowe, P, Munton, R, and Flynn, A 1993, *Constructing the Countryside*. UCL Press, London.

Marsden, T, Yu, L, and Flynn, A 2011, "Exploring ecological modernization and urban-rural ecodevelopments in China: The case of Anji county," *Town Planning Review*, vol. 82, no. 2, pp. 195–224.

Milestad, R, Bartel-Kratochvil, R, Leitner, H, and Axmann, P 2010, "Being close: The quality of social relationships in a local organic cereal and bread network in Lower Austria," *Journal of Rural Studies*, vol. 26, no. 3, pp. 228–240.

Ortiz-Miranda, D, Moreno-Pérez, O and Moragues-Faus, A 2010, "Innovative strategies of agricultural cooperatives in the framework of the new rural development paradigms: The case of the Region of Valencia (Spain)," *Environment and Planning A*, vol. 42, no. 3, pp. 661–677.

Paldam, M and Svendsen, GT 2001, "Missing social capital and the transition in Eastern Europe," *Journal for Institutional Innovation, Development and Transition*, vol. 5, pp. 21–34.

Qiao, Y, Martin, F, Cook, S, He, X, Halberg, N, Scott, S, and Pan, X 2018, "Certified organic agriculture as an alternative livelihood strategy for small-scale farmers in China: A case study in Wanzai county, Jiangxi province," *Ecological Economics*, no. 145, pp. 301–307.

Renting, H, Marsden, TK, and Banks, J 2003, "Understanding alternative food networks: short food supply chains and rural development," *Environment and Planning A*, vol. 35, no. 3, pp. 393–411.

Renting, H, Rossing, WA, Groot, JC, van der Ploeg, JD, Laurent, C, Perraud, D, Stobbelaar, DJ, and Van Ittersum, MK 2009, "Exploring multifunctional agriculture. A review of conceptual approaches and prospects for an integrative transitional framework," *Journal of Environmental Management*, vol. 90, no. 2, pp. 12–123.

Sharpley, R 2002, "Rural tourism and the challenges of tourism diversification: The case of Cyprus," *Tourism Management*, vol. 23, no. 3, pp. 233–244.

Song, Y, Qi, G, Zhang, Y, and Vernooy, R 2014, "Farmer cooperatives in China: Diverse pathways to sustainable rural development," *International Journal of Agricultural Sustainability*, vol. 12, no. 2, pp. 95–108.

Suh, J 2015, "Communitarian cooperative organic rice farming in Hongdong district, South Korea," *Journal of Rural Studies*, vol. 37, pp. 29–37.

Sultan, T and Larsén, K 2011, "Do Institutional incentives matter for farmers to join cooperatives: A comparison of two Chinese regions," *Journal of Rural Cooperation*, vol. 39, no. 1, pp. 1–18.

Tisenkopfs, T, Kovách, I, Lošťák, M, and Šūmane, S 2010, "Rebuilding and failing collectivity: Specific challenges for collective farmers marketing initiatives in post-socialist countries," *International Journal of Sociology of Agriculture and Food*, vol. 18, no. 1, pp. 70–88.

United Nations 2001, *Cooperatives in Social Development, Report of the Secretary General*. Accessed at www.un.org/documents/ecosoc/docs/2001/e2001–68.pdf on May 6, 2018.

van der Ploeg, J 2000, "Revitalizing agriculture: farming economically as starting ground for rural development," *Sociologia Ruralis*, vol. 40, no. 4, pp. 497–511.

van der Ploeg, J, Roep, D, Renting, H, and Banks, J 2002, "The socio-economic impact of rural development processes within Europe." In: van der Ploeg, J, Banks, J, and Long, A (eds.), *Living Countrysides. Rural Development Processes in Europe: The State of the Art*. Elsevier, Doetinchem.

van der Ploeg, J, Laurent, C, Blondeau, F, and Bonnafous, P 2009, "Farm diversity, classification schemes and multifunctionality," *Journal of Environmental Management*, vol. 90, pp. 124–131.

van der Ploeg JD, Long N, and Banks, J, 2002, *Living Countryside. Rural Development Processes in Europe: The State of the Art*. Doetinchem: Elsevier.

van der Ploeg, J, Renting, H, Brunori, G, Knickel, K, Mannion, J, Marsden, T, de Roest, K, Sevilla- Guzmán, E, and Ventura, F 2000, "Rural development: From practices and policies towards theory," *Sociologia Ruralis*, vol. 40, no. 4, pp. 391–408.

van der Ploeg, J, Ye, Z, and Schneider, S 2012, "Rural development through the construction of new, nested, markets: Comparative perspectives from China, Brazil and the European Union," *The Journal of Peasant Studies*, vol. 39, no. 1, pp. 133–173.

Wen, T and Dong, X 2010, "Village-community rationality: A new perspective to solve the predicaments of 'sannong' and 'sanzhi,'" *Journal of the Party School of the Central Committee of the CPC*, vol. 14, no. 4, pp. 20–23 (in Chinese).

Wilson, G 2010, "Multifunctional 'quality' and rural community resilience," *Transactions of the Institute of British Geographers* New Series, vol. 35, no. 3, pp. 364–381.

Yan, H and Chen, Y 2013, "Debating the rural cooperative movement in China, the past and the present," *The Journal of Peasant Studies*, vol. 40, no. 6, pp. 955–981.

Yan, H and Chen, Y 2015, "Agrarian capitalization without capitalism? Capitalist dynamics from above and below in China," *Journal of Agrarian Change*, vol. 15, no. 3, pp. 366–391.

Yang, H, Vernooy, R, and Leeuwis, C 2017, "Farmer cooperatives and the changing agri-food system in China," *China Information*.

Yang, S and Wen, T 2011, "The predicament of and solution to farmers' organization," *People's Forum*, vol. 29, pp. 44–45 (in Chinese).

Zhang, X 1999, "Cooperatives, communities and the state: The recent development of Chinese rural cooperatives in transition," *Land Reform, Land Settlement and Cooperatives*, vol. 1–2, pp. 92–105. Accessed at ftp://ftp.fao.org/sd/SDA/SDAA/Lr99/X3720T08.pdf.

5 Bottom-up initiatives

The emergence of "alternative" food networks

Zhenzhong Si, Theresa Schumilas, and Steffanie Scott

Introduction

Chapter 2 of this book has reviewed the rapid developments of China's food system concerning the industrialization of agriculture, consolidation of food production and processing, supermarketization of food retailing, and changing patterns of food consumption. The same trend has been observed in many other countries where the construction, implications, and evolution of alternative food networks (AFNs) or alternative systems of food provision (Watts et al. 2005) have attracted a great deal of scholarly attention since the mid-1990s (e.g. Goodman 2003, 2004; Maye, Kneafsey, and Holloway 2007; Tregear 2011). AFNs are "rooted in particular places, and they aim to be economically viable for farmers and consumers, use ecologically sound production and distribution practices, and enhance social equity and democracy for all members of the community" (Feenstra 1997: 2). AFNs proliferate as reflexive responses to the industrialization of the food sector but also face "mainstreaming" challenges (see Goodman, DePuis, and Goodman 2012). Types of AFNs include community-supported agriculture (CSA) (Feagan and Henderson 2009; Lang 2010), farmers' markets (Kirwan 2004, 2006; Brown and Miller 2008; Smithers, Lamarche, and Joseph 2008; Beckie, Kennedy, and Wittman 2012), buying clubs (Little, Maye, and Ilbery 2010) public procurement programs (Allen and Guthman 2006; Kirwan and Foster 2007), community gardens, and more (see Goodman and Goodman 2008; Tregear 2011; Raynolds 2000). The main (and most well-known) AFN "civic organizations" and initiatives are those in the UK, other parts of Western Europe, and North America. In contrast, initiatives in emerging economies tend to be more recent and have received less recognition (but see Wang et al. 2015; Krul and Ho 2017; Escher, Schneider, and Ye 2018; Ding, Liu, and Ravenscroft 2018; Abrahams 2007; Rocha and Lessa 2009; Freidberg and Goldstein 2011; Shi et al. 2011a; Scott et al. 2014).

As the world's second largest economy and largest developing country, China is experiencing rapid growth in food production and consumption as well as fundamental transformations in its food system. From a country that struggled with food sufficiency to a country immersed in food safety crises in recent years, China is gradually transforming its food system from a state-coordinated food security–oriented one to a system with nascent but increasing civil society and private-sector participation (Scott et al. 2014). Meanwhile, food activists in China are adapting alternative food production and provisioning initiatives from North America and Europe, including organic production, CSAs, farmers' markets, and buying clubs. Some other endogenous initiatives such as "weekend farming" are also thriving (Fei Liu 2012). However, although a small number of studies have addressed the organic and ecological agriculture sector in China (see Shi 2002; Thiers 2002, 2005; Ye, Wang, and Li 2002; Shi and Gill 2005; Sanders 2000, 2006; Sheng et al. 2009; Qiao 2010), AFNs such as CSAs, farmers' markets and buying clubs have received less scholarly attention (but see Shi et al. 2011a; Scott et al. 2014; Krul and Ho 2017). Likewise, there have been few studies of these civil society initiatives in other emerging economies. The absence of Chinese AFNs in agrofood literature is partly due to the fact that AFNs were conceptualized within a Western context, but also because most of them have only emerged in China since 2008.

In response to these gaps, this chapter seeks to understand the characteristics of AFNs in China, the various conditions shaping them, and how they differ from AFNs in the West. To characterize AFNs in China, we first develop a typology of food initiatives in China that would usually be categorized as AFNs in the West. We then illustrate their emergence and operations with specific cases by probing into their key principles, inherent values, and internal contradictions to examine their alternativeness. We argue that it is the elements of alternativeness emphasized in these initiatives that distinguish them from Western ones. We conclude the chapter with a discussion of the economic, ecological, political, and cultural conditions that shape these initiatives.

Dimensions of alternativeness within alternative food networks

Among the various facets of AFNs that have captured the attention of agrofood scholars, one intriguing issue is the interrogation of "alternativeness." Indeed, the alternativeness of AFNs should not be taken for granted. Rather, its existence and characterization should be examined in specific socio-economic and political contexts. While the dichotomous characterization of food venues as "alternative" versus "conventional" may seem too simplistic and problematic (Sonnino and Marsden 2006), food initiatives such as CSAs, farmers' markets, and buying clubs still possess particular attributes that distinguish them, to various extents, from mainstream market venues and thus underpin their alternativeness. According to Whatmore, Stassart, and Renting (2003: 389), these novel initiatives are generally conceptualized under the AFN umbrella based on three main dimensions of "alternativeness" that they have in common:

1   their constitution as/of food markets that redistribute value through the network against the logic of bulk commodity production;
2   they reconvene "trust" between food producers and consumers; and
3   they articulate new forms of political association and market governance.

While these representations and appeals of AFNs in the economic, social, and political spheres characterize most of their fundamental features, we would add a fourth dimension—ecological alternativeness (see Jones et al. 2010). Ecological alternativeness addresses a common feature of many AFNs to the extent that they embrace ecological production practices. The ecological production practices of these AFNs drove us to study them as bottom-up initiatives to develop the ecological agriculture sector. We argue that the ecological alternativeness alongside the other three major dimensions constitute the fundamental AFN discourses, and underpin AFNs' tension with the hegemonic industrial food system.

The first dimension of alternativeness identified by Whatmore, Stassart, and Renting (2003) concerns the redistribution of value to smallholders along the value chain. Alternative and local networks generally have the goal of improving economic viability of local farms by providing stable local markets and shortening value chains (Allen et al. 2003). The sentiment of going against "the logic of bulk commodity production" (Whatmore, Stassart, and Renting 2003: 389) in AFNs is mirrored in the promotion of CSAs, farmers' markets, and small-scale independent farms. Although empirical studies reveal that AFNs may not guarantee local and small producers more profit (Brown and Miller 2008; Goodman 2009), the alternativeness of value redistribution is such a strong emphasis among food activists that Allen (2010: 300) suggests that American agrarianism, which upholds "the moral and economic primacy of farming," results in an emphasis on improving the viability of the family farm over social justice concerns.

The second dimension of alternativeness of AFNs is the reconnection between producers and consumers. Alternative food discourses highlight local modes of production and distribution (Allen et al. 2003; Feagan 2007) and direct encounters that reconnect consumers and producers (Holloway et al. 2006; Wiskerke 2009). The face-to-face interaction in AFNs conveys relationships that are more than impersonal commodity exchanges but rather a connectivity that embodies a personalized "sentiment of regard" (Kirwan 2004). The sentiment-infused "social ties, personal connections, and community good will" define the social embeddedness of AFNs (Hinrichs 2000: 301). Correspondingly, reconnection between producers and consumers, alongside "re-placing" and "relocalization," are seen by agrofood scholars as some of the most prominent features of alternative food initiatives (Kirwan 2004; Watts et al. 2005; Wiskerke 2009). This suggests that reciprocity, rather than the dominance of either consumer or producer, defines the reconnection. Consequently, this understanding of reconnection leads to the specific focus of "trust" within the local agrofood networks literature (see Jarosz 2000). The political economy perspective of AFNs studies sees the local as a site of resistance and, in emphasizing spatial relations, is concerned with the micro-politics of place and relations of trust and reciprocity. Reconnection and trust are seen as inherent components of alternativeness in AFN discourses.

The third dimension of alternativeness relates to the seeking of new forms of food governance and political agendas, such as the thriving nongovernmental food organizations and associations (e.g. Toronto Food Policy Council). AFNs are believed to have the potential to alter the current institutional arrangements for food provisioning. Some researchers (Lyson 2004; Alkon 2008) have also pointed out that sustainable agriculture and consumption have the potential to "reinvigorate democracy." Scholars have also explored the possibility of new food policies such as inscribing institutional food procurement into public policy (Allen and Guthman 2006). The political dimension of alternativeness has significant implications for food politics, an arena where various players struggle to reconfigure food production, consumption, and regulation (e.g. Nestle 2007).

The ecological nature of alternative food initiatives (Allen et al. 2003, Marsden and Smith 2005) is the fourth dimension of AFNs' alternativeness we want to highlight here. It is particularly relevant to organic and other forms of ecological production practices (see Scialabba and Müller-Lindenlauf 2010). It is also a prominent issue as reflected in AFNs' claim of reducing greenhouse gas emissions by reducing "food miles" and carbon footprints involved in long-distance food transport. The ecological dimension is also mirrored in many AFNs' promotion of eating local, seasonal, and plant-based diets, as opposed to out-of-season and animal-based foods sourced from global markets (see Feenstra 1997; Jarosz 2008). In this way, nature, whose importance is continuously being outflanked or reduced in the industrialized food system (Murdoch, Marsden, and Banks 2000), has been extensively integrated in a more positive manner into AFNs.

Despite the diverse dimensions of alternativeness within AFNs, we argue that there has been insufficient consideration of the extent to which all of these dimensions apply across AFNs in different contexts. As Jarosz (2008: 242) noted, "AFNs are not static objects ... they emerge from political, cultural, and historical processes." In specific political economies such as China, the full spectrum of alternativeness in AFNs is not necessarily as present as elsewhere. Rather, the manifestations of these dimensions, which comprise the dynamic landscape of AFNs, are context-specific. Indeed, our research in China suggests that the manifestation of alternativeness of AFNs varies in different economic, social, and political contexts. For example, because of the fewer chemical inputs in alternative food production (particularly ecological and organic agriculture) and in less processed food, there is a general assumption of the healthfulness of food in AFNs. For health reasons—to reduce their exposure to agrochemicals and to antibiotics in meat—Chinese consumers are seeking out organic and ecologically produced foods via alternative food procurement channels (see Shi et al. 2011a; Scott et al. 2014). However, discussions about alternativeness in AFN literature have paid much less attention to this "healthfulness" element.

Critical studies of AFNs in North America and Europe question the various dimensions of AFNs' alternativeness, particularly regarding their claims of social inclusion (Hinrichs and Kremer 2002; Guthman 2008), social justice (Hinrichs and Allen 2008; Allen 2010; DeLind 2011), and environmental outcomes (Hinrichs 2003; DuPuis and Gillon 2009; Jones et al. 2010). These critiques of alternativeness further raise concerns that the current conceptualization of alternativeness have not sufficiently recognized the variation in dimensions of alternativeness across diverse social, political, and economic contexts.

Therefore, to overcome the critiques of the binary view of alternative versus conventional characterizations of food systems, we argue that a further unpacking of the existing four dimensions of alternativeness is necessary. This will not only address the concern of over-simplification in examining alternativeness but also enable a more operable analytical framework for characterizing AFNs in diverse contexts. Based on the four major dimensions of alternativeness identified in the previous section, and taking our interview results into account, we further unpack the dimensions of alternativeness embedded in AFNs into eight elements, namely: healthy, ecological, local, seasonal, small-scale, strengthening social ties and personal connections, socially just, and political. These elements are projections from the four major attributes of food and various relations embedded within AFNs.

Drawing on Ho and Edmonds's (2007) conceptualization of China's "embedded activism," we argue that the current AFNs in China are strongly situated in the country's political economy. These emerging alternative food initiatives have a strongly shaped alternativeness that is embedded within, and is also a reflection of, local geographies. In other words, AFNs in China display strong evidence of alternativeness around food "healthfulness" and nutrition, but weak representations of social and political elements in terms of reconnection, social justice, and forms of political association.

The typology and characteristics of alternative food networks in China

As the common saying in the West goes, "you are what you eat." Traditional Chinese culture believe "food is Heaven," suggesting that food is sacred and central in traditional Chinese culture. Yet, this belief has been shattered by numerous food safety scandals in recent years (Pei et al. 2011; Yang 2013; Klein 2013); food is no longer an innocent and dignified sphere of people's lives. However, rather than a "retreat of the state to baseline food safety regulation," as has happened in many advanced economies (Goodman, Dupris, and Goodman 2012: 88), the state in China has taken a more proactive role to promote quality food production, issuing a set of national quality food standards for not only organic but also "green" and "hazard-free" food (see Chapter 3). To cope with the widespread distrust of organic certification due to frequent reports of fraudulent organic products in markets (Yin and Zhou 2012), the state enacted a much more—some would say overly—stringent organic standard in 2012 (Scott et al. 2014).

Another important change that has had profound implications for the emergence of AFNs is the growing purchasing power of the middle class (Shi et al. 2011a). A recent report categorized 53 per cent of Chinese population into lower-middle class and 7 per cent of them into upper-middle class based on annual disposable income per capita in 2015 (Leung 2016). Barton, Chen, and Jin (2013) estimated that 75 per cent of Chinese urban consumers would be considered middle-class by 2022, in terms of their purchasing power. A characterization of the shareholders in the most well-known CSA in China—Little Donkey Farm in Beijing—revealed that, compared to poorer segments of the population, the middle class has a stronger interest in quality food and in multifunctional urban agriculture that integrates food production and recreational functions (Shi et al. 2011b).

The mounting food safety crisis and the growing middle class, alongside other factors, has propelled Chinese civil society actors to initiate various alternative food ventures in Chinese cities since about 2008. These initiatives mainly include CSAs, ecological farmers' markets, buying clubs, and urban people engaging in self-provisioning through recreational gardening through rental plots. Except for the recreational garden plot rentals, these emerging initiatives in China were not endogenous but introduced from North America and Europe. However, they are significantly different from their Western counterparts in terms of the manifestation of alternativeness. This is evidenced by, for instance, producer–consumer reconnection. The reconnection in many Chinese AFNs is more narrowly built upon the food safety appeal and not genuine connections with those who are hired to do the farm work. This odd situation is linked to the general negative perceptions of peasants—typically portrayed as selfish and shortsighted.[1]

Based on the four major dimensions of alternativeness identified in the previous section and our analysis of Chinese AFNs, we further unpacked the four dimensions of alternativeness embedded in AFNs into eight elements (see Table 5.1). These elements pertain to either the features of food within these AFNs or to the relationships among stakeholders (between producers and consumers, producers and nature, and among producers themselves). We also identified alternative food initiatives that reflected these elements, as well as the connections between these elements and consumer motivations. This analysis underscores how the "situated AFNs" in China reflect a very different landscape of alternativeness from those in the West.

Our cases of CSAs, farmers' markets, buying clubs, and recreational garden plot rentals demonstrate different elements of alternativeness from the perspectives of their organizers. In contrast to the diverse ethical values represented among the organizers, consumers tended to have a single focus on healthfulness of food. Although the ecological and health elements are intertwined, the main motivation of consumers seemed to be individualistic health concerns, rather than a broader environmental ethic. In fact, we found a lack of ecological concern among Chinese consumers in general, even when ecological alternativeness is a characteristic of the food sold at these venues. Being local is another imperative feature of AFNs that shapes the alternative food movement in the West but is also noticeably weak among the motivations of Chinese consumers, although some CSAs, farmers' markets, and buying clubs do promote "eating local." Other elements of alternativeness are still at the early stage of being communicated by food activists to patrons of alternative food initiatives. The following section examined the alternativeness of these four major types of alternative food distribution networks in China with specific cases.

Community-supported agriculture farms

Although there has been no accurate data about the exact number of CSA farms in China, some estimated it to be around 500 (Merrifield and Shi n.d.). One of the difficulties in estimating the number of CSAs in China is that this term is being widely used as a branding or marketing label. We found that many online ordering stores refer to themselves using the English acronym CSA. So, while this might describe a small-scale farmer who enrols members in a way quite similar to the understanding of this term in the West, it could also describe a much larger business that aggregates product, which may or may not be ecologically produced, from multiple farms, and makes this available through quite sophisticated online storefront operations. Given this confusing landscape, we considered CSAs to be initiatives in which an operator (either a peasant farmer or an urban resident) sells products from farmland that they themselves manage, to an established group of buyers who signed up to be the members of the farm. Thus, we excluded cases where consumers ordered online from a list of options without direct contact with the CSA operator, and with no possibility of visiting the farm.

Table 5.1 Unpacking the alternativeness of AFNs in China

Types of alternativeness	Elements of alternativeness	Representative AFN initiatives in China				Consumer motivations for each element
		CSAs	Farmers' Markets	Buying Clubs	Recreational garden plots rentals	
Food features	Healthy (free from chemical residues and more nutritious)	✓	✓	✓	✓	Strong
	Ecological	✓	✓			Relatively weak
	Local	✓	✓	✓		Relatively weak
	Seasonal	✓	✓			Weak
Relationships among stakeholders	Small-scale*		✓			Weak
	Social ties and personal connections	✓	✓	✓	✓	Weak
	Social justice*		✓	✓		Weak
	Political*a		✓			Weak

Notes
* These elements were rarely mentioned by our interviewees.
a Political refers to the AFNs' alternativeness in "articulating new forms of political association and market governance" (see Whatmore et al. 2003: 389).

The CSA model examines in our study was introduced from North America by a group of well-educated activists and farmer entrepreneurs, who also integrated traditional practices of sustainable farming into its practice. The first CSAs in China were CSA farms in Anlong village in Chengdu, Sichuan province (established in 2006), and Little Donkey Farm in Beijing (established in 2008). As an illustration of the rapid growth of CSAs in recent years, China held its ninth annual symposium in December 2017, which attracted more than 1,000 attendants, including CSA operators, researchers, social organizations, and farmers from across the country.

Chinese CSAs often adopt organic farming practices, which distinguish them from conventional farms. In contrast to organic farms established by private enterprises examined in Chapter 3, CSA farm owners in most cases choose not to seek organic certification, because of not only the prohibitive cost but also the poor reputation of organic certification (Yin and Zhou 2012; Wang et al. 2015). Many farms instead prefer to develop a loyal customer base through farmers' markets, word of mouth, and personal relations. Customers are invited to visit their farms and ask questions. This is sometimes referred as "participatory certification" or "ethical inspections" (*liangxin renzheng* in Chinese). This entails customers hearing farmers' promises and descriptions of their practices, inspecting the farming practices for themselves, and then deciding whether to buy the food or become a shareholder of the farm.

The introduction of CSAs and some ecological farms in China exemplify a nascent values-based movement to promote consumer–producer and urban–rural connections (see Paüla and McKenzie 2013; Krul and Ho 2017).

Figure 5.1 Steffanie beside a greenhouse at Little Donkey Farm in Beijing.

A group of Chinese academic researchers has contributed to the development of CSAs in various ways, including as advocates for the establishment of organic farms and as consultants to local and central governments. China Renmin University in Beijing, through the leadership of Professor Wen Tiejun, has been particularly noteworthy in the promotion of CSAs, peasant cooperatives, and the social economy (Shi et al. 2011a; Wen et al. 2012; Pan and Du 2011a, 2011b).[2] NGOs, though few in number and constrained to some extent in China, have also been an important catalyst (Ju 2009). The Hong Kong–based Partnerships for Community Development (PCD) is one of the most active NGOs in supporting CSA development in China. It worked with the Chengdu Urban Rivers Association (a local NGO) to help establish the CSAs in Anlong village, Sichuan province. It also facilitated the establishment of many other CSAs in southern China.

While each CSA has its own unique mix of products and pricing, and they vary in size and managerial structure, they should not be considered atomistic farms. Indeed, these operations are interconnected in multiple ways. During our fieldtrips, we were continually surprised at the strength of the connections among and between CSA operators, farmers' markets, and buying clubs in China, especially considering that many of these AFNs had only recently been established. Each individual operator routinely referenced the others and dozens of media articles and online directories listed the same group of networked farms, markets, and buying clubs. Many of the same farm operators were present at meetings and conferences we attended.

How "alternative" were these CSAs in terms of the eight elements identified above? Our interviews with CSA farmers and their interns revealed a strong understanding of the ecological alternativeness and its health implications. CSA farmers believed that avoiding chemical fertilizers and pesticides would contribute substantially to environmental sustainability. The slogan "eat local, eat seasonal" was promoted by a small number of food activists and some CSA farms. "Social ties and personal connections" among CSA farmers and between CSA farmers and their customers were also highly valued (see Table 5.1).

Despite some evidence of these elements of alternativeness, our fieldwork showed that the degree of their alternativeness is open to question. As many of the CSA farms in China are founded by market-oriented entrepreneurs, operating within rather than beyond the neo-liberal market logics, it is hard for them to escape the circle of profit-motivated commodity production. Some of the elements of alternativeness may thus be subdued in order to cater to consumer needs. For example, although "eating seasonally" has been widely praised by CSA farmers, we still observed an online debate on microblogs between some CSA farms on whether it was appropriate to grow vegetables in greenhouses, as this was seen by some as violating the principle of "eating seasonally." As noted earlier, procuring safe food is the main motivation for consumers to participate in Chinese CSAs (Ju 2009; Gale 2011). Therefore, neither community building via producer–consumer reconnection nor value redistribution to small producers is a key priority in many CSAs, although they are priorities for some CSA operators.[3] Social justice as an element of alternativeness is not reflected here. In fact, we observed a strong feature of "elite capture" in the class and racial complexion of CSAs: the dominance of well-educated farm operators noticeably excludes real peasants in decision-making.[4] Peasant farmers who hold the original land contractual and operational rights on the farmland are often hired as farm workers but their opinions were often not welcome. For their part, CSA shareholders preferred to interact with farm managers or owners (urban entrepreneurs called "new farmers" with secondary degrees (Fang Liu 2012)). Thus, small-scale farmers are not empowered, nor is their social status boosted.

Recognizing this lack of attention to social justice concerns and the exclusion of peasant farmers, a small group of Chinese food activists initiated a new CSA in 2012 in Tongzhou district in Beijing—Shared Harvest Farm—to experiment with value redistribution through the model of working with, rather than hiring as labour, small peasants, and "sharing more harvest" with them. However, owing to various challenges they encountered in working directly with peasants, they returned to the original model on their second production base in Shunyi district. Moreover, as a result of enormous private capital penetration in organic agriculture in the last few years (Yuan 2011), many farms have been co-opting the term "CSA" and instrumentally using it as a marketing buzzword, with little attention paid to ecological sustainability or risk sharing. In fact, much of China's organic production has been subsumed by large food companies and operated in the same way as a conventional food business.[5]

The political element in terms of articulating "new forms of political association and market governance" is also minimal among CSA farmers in China. It is noteworthy, nevertheless, that the Rural Reconstruction Center at Renmin University in Beijing has been holding annual nationwide CSA symposiums since 2009. At the 2012 symposium, CSA farmers decided to establish a National Ecological Agriculture Cooperation Network, aiming at sharing information and knowledge. In November 2015, at the Sixth International Symposium on Community Supported Agriculture, held in Beijing, the network transformed into China Social Ecological Agriculture CSA Alliance and was officially registered as a social organization in 2017. However, it is still not clear how this initiative will be translated into a new form of political association and market governance.

Ecological farmers' markets

Another noteworthy type of AFNs in China is farmers' markets.[6] In several large cities, including Beijing, Shanghai, Guangzhou, Tianjin, Xi'an, Chengdu, Nanjing, Chongqing, Kunming, and Guiyang, organic (sometimes called green or ecological) farmers' markets have become a new alternative food venue that attracts large numbers of middle-class consumers. In contrast to traditional "wet markets," where petty-traders bring products from large wholesale markets to smaller urban markets for resale, in the farmers' markets we studied, farmers themselves sell directly to urban consumers. These ecological farmers markets, most of which emerged between 2009 and 2010, aim to rebuild the trust between consumers/eaters and food producers and serve as a platform for education and advocacy.

The Beijing Farmers' Market, formerly named "Beijing Country Fair," is the most prominent example of these new-style organic farmers markets. In 2017, the market was operated by five full-time employees and a group of volunteers. They sometimes also organized public talks for followers of their microblog (which numbered more than 118,500 in July 2017) and for subscribers of their WeChat account. The inspiration of the major founder came from her experience in New York's farmers' markets (Shu 2012). To afford the fees associated with operating the market, the market collected a small fee from vendors, received some grants from NGOs to cover the staffs' salaries, and also earned income from the "Country Fair Kitchen" by selling food at the market prepared using produce from the market.

In 2017, the market was held two or three times a week in different locations in order to be accessible to people in various parts of the city. Market locations have included department stores, campuses, residential

Figure 5.2 Steffanie talking to a farmer at the Beijing Organic Farmers' Market.

neighbourhoods, hotels, and shopping malls. The time and location were publicized on the market's microblog and their WeChat account each week. More than 20 vendors (out of the 40 approved farms, NGOs, social enterprises, and other merchants) turned up regularly. Goods sold at the market were mainly fresh produce and prepared foods (tofu, rice wine, baked goods, cheese), plus occasionally handicrafts such as soap. Although the prices there were several times higher than conventional food, products would often sell out quickly.

An interrogation of the farmers' market according to the eight elements listed in Table 5.1 shows that the market demonstrates all these elements, many of which are manifested in the criteria for selecting vendors. Most farms selling goods at the market were not certified organic but were screened through informal "inspections" by the organizers based on the following criteria: they are small- or medium-scale, use no synthetic fertilizers or pesticides, animals are not caged and no unnecessary antibiotics are used, and farmers are willing to work with others to develop the market.[7] These gatekeeping rules helped the market to maintain a high reputation compared to certified organic food sold in supermarkets and other outlets.

In recent years, the organizers have facilitated a participatory guarantee system (PGS) for peer certification of these farmers, to take the onus off organizers for conducting the inspections. PGS, adopted in a growing number of countries, uses participatory monitoring to maintain the organic status and reputation of the whole group (Nelson et al. 2010). Farmers, market managers, and customers formed inspection committees to visit and informally inspect farms. In 2016 alone, the market organized more than 200 farm visits. It demonstrates a type of new association among various stakeholders involved. Two of the market organizers that we interviewed also expressed their serious concerns about the industrialized food system and their wishes to restructure it. In addition, the market claims to be a "place to foster connections between farmers and consumers" (Beijing Farmers' Market 2011), where a sense of community is forged and developed. Thus, all elements of alternativeness are represented in this farmers' market, although to different extents.

Despite these elements of alternativeness being represented in this farmers' market, many are only perceived by the market's key facilitators, not by its ordinary customers. Rather, it is food safety and food quality concerns that attract most consumers here (Shu 2012). We observed that customers of the market, who came from every corner of the city, were generally white-collar workers, expectant mothers and mothers of young children, or elderly people with poor health conditions. These groups are believed to have the strongest demand for healthy food.[8] Thus, the loyalty of consumers at these markets is typically based on their trust in the safety and quality of the food, rather than a deeper interest in connecting with producers. The market manager expressed her concern about the difference in values between market organizers and customers:

> For us [market organizers], being ethical and giving attention to social justice are the most important criteria. After that we are concerned that the products are organic, local, and small-scale. But we also know we need to keep diversifying to make the market attractive to a broad group of consumers ... the healthfulness of food is a window to attract consumers. Although I want to promote the values of farmers' markets to ordinary customers, I don't want to scare them away.
>
> (Interview with one market organizer in Beijing, December 6, 2012)

Despite its strong ethical positions, the Beijing Farmers' Market organizers face criticism from customers for being too "producer-centred" and "disparaging consumers' interests" by emphasizing the central position of farmers within producer–consumer relations, giving farmers a role as educators of consumers.[9] This poses a threat to the reconnection between farmers and consumers. In some of the market venues, the overcrowded and busy market offers little time or space for direct communication, which then diminishes the scope for building mutual "trust" and makes it merely a venue for direct-to-consumer marketing (see Zhang 2013).

The market also faces critiques from those who disagree with its use of the term "organic" in its promotion. This touches upon a critical debate within organic food production in China: whether producers should get organic certification or not. In response to the critiques, the market organizer explained that,

> In China, the term "organic" has been "polluted." We want to bring back its true meanings. Many people believe that "organic" is a result of certification and always want to compare to the standards [when judging whether a certain type of food is "organic"], but we believe "organic" is an idea that means farming sustainably and reducing the environmental cost.
>
> (Interview with one market organizer in Beijing, December 6, 2012)

This debate over certification reveals the complexities and competition surrounding AFN language in China, which deserves further analysis. The struggle over appropriating "organic" language could severely affect its legitimacy in competing for alternative economic space.[10] Consequently, it will affect the way that the alternativeness of the Beijing Farmers' Market and many small-scale farms are represented.[11]

Buying clubs

Buying clubs are another strong consumer-driven initiative that responds to the widespread food safety anxiety in China. The earliest buying club known in China emerged around 2004, when a group of self-described "nature lovers" started to regularly purchase homegrown produce from nearby farmers in Liuzhou city, in Guangxi Zhuang Nationality Autonomous Region in southwest China. Later, housewives and a group of volunteers in Beijing and Shanghai facilitated their own buying clubs driven by strong concerns about food safety. Well-known buying clubs include *Ainonghui* (Care for Farming Group) established in Liuzhou in 2004, Green League, which started in Beijing in 2010, Shanghai *Caituan* (Group Procurement of Vegetables), which began in 2010 in Shanghai, and Green Heartland, founded in Chengdu in 2010.

Green Heartland in Chengdu city, Sichuan province, is one of the most prominent cases. Its activities date back to 2007 when a group of urban residents got to know the first CSA farmers in China. A local NGO, Chengdu Urban Rivers Association, supported by a Hong Kong–based NGO, Partnerships for Community Development, introduced the residents to the farmers in Anlong village near Chengdu. They gradually formed a consumer group. Their activities went beyond group procurement of healthy food to also include organizing a periodic farmers' market at a local market, arranging for their members to visit farms, providing members with opportunities to experience farming, and educating farming knowledge. It was not only a way of informal inspection (which they call "conscience certification," to ensure that their suppliers were farming in a sustainable way), but also a process of building closer relations. They brought together farmers in Sichuan province by organizing a farmers' market. At least 10 per cent of their sales were donated to buy food for poor families in a local community. They also collected a small fund for their activities by selling homemade jam and soap.[12]

Figure 5.3 Zhenzhong (second from left) and Steffanie (in the middle) with founders of the Green Heartland buying club in Chengdu, Sichuan province.

When examining the alternativeness of buying clubs in China, we found that they were initiated entirely by informed middle-class consumers with a strong concern about healthy and safe food. Similar in profile to those who procure food via CSAs and farmers' markets, their major motivation was to have access to safe and healthy food, usually to foster their children's health. This is reflected in the unique characteristics of the people—housewives with children—who founded several major buying clubs in China.[13] Their desire to purchase from local farmers and traditional farmers in remote areas so that these farmers can get a decent compensation for their products demonstrates a certain level of alternativeness in "local" and "social justice" elements.[14] Activities organized by these buying clubs for their members also demonstrate a concern over "social ties and personal connections." However, other elements were absent.

Compared to CSAs, the number of buying clubs in China is much smaller. Hence, despite the strong ethical values that the Green Heartland buying club holds, it is hard to judge whether buying clubs emerging in the future would promote these principles to the same extent. In addition, it is a huge challenge for the small number of initiators to effectively communicate their ethical values to the rapidly growing number of members, whose primary motivation for joining the buying club is simply to have access to safe and healthy food.

Recreational garden plot rentals

Renting a plot (known as "rental farming" or "weekend farming") in peri-urban areas is another type of alternative food initiative. In this type, consumers engage more directly in food production. Since 2009, many ecological farms (usually CSAs) in peri-urban areas have begun to rent out small plots (e.g. $30 \, m^2$ in size) and provide advice to urbanites who are keen to grow their own organic vegetables (similar to community garden plots in North America). These urbanites usually proudly call themselves "weekend farmers" or "mini landlords" (Little Donkey Farm 2012). They visit their plots at least once every weekend. The cost of renting a plot near large cities in China is usually a few thousand Chinese yuan (US$500 or more) per year, many times higher than getting access to a community garden plot in the West. It is mainly provided by farms located close to urban centres. One explanation for the emergence of garden plot rentals is the extreme popularity, between 2009 and 2013 among Chinese white-collar workers, of Happy Farm, an online social network game for multiple players. It allows players to virtually grow and harvest their own crops, trade them with others, and even steal from neighbours (China Agriculture Information Web 2013).

What is alternative about these plot rentals? A close examination of the experiences of these "mini landlords" (Little Donkey Farm 2012) reveals that there are sophisticated physical, mental, and philosophical motivations

Figure 5.4 A plot rented by Air China for its employees at Anlong village near Chengdu, Sichuan province.

that inspired these "weekend farmers." These include food safety concerns (to save themselves from the food safety crisis), leisure (escaping busy city life), affinity with nature, physical exercise, and the emotional needs of seniors who live with their family in the city. With similar sentiments to the "back-to-the-land" movement in the West, seniors in urban families find renting plots a good way to relive their nostalgia for the old times, educate their children in the countryside, feel a sense of belonging, communicate with friends, and enact their values. However, besides the healthfulness of food and building social ties with others (not necessarily farmers on the farm), other elements of alternativeness are largely absent here.

Compared to other types of AFNs, recreational garden plots in China are an AFN that is more fully embedded within the Chinese social and political context, rather than being a model imported and adapted from elsewhere.[15] This context can be understood in terms of three different elements. First, the emergence of plot renting is a direct response to the severe food safety crisis in an environment of an extreme lack of trust of food producers and processors. Responses from a diverse range of interviewees reinforced this point. Second, the form of plot renting that entails renting a small piece of land is also linked to the collective but scattered land rights system. Renting is the only option for urbanites who want to farm but are not allowed to purchase land from collective land owners. As we explained in previous chapters, under the Household Responsibility System in China, farmland within an administrative village is distributed among its collective village members. This imposes a great challenge for CSA operators to acquire consolidated areas of farmland. Accordingly, plots rented to urbanites are small. Third, the popularity of rental farming among urbanites also reflects the social problems associated with rapid urbanization in China. Many renters are looking for a plot for their elderly parents, who had been farmers for their whole lives but had moved to the city to live with their children, many of whom are the first migrant worker generation in cities. Detachment from land leads to an "emotional vacuum" for the seniors trying to fit into city life (Little Donkey Farm 2012). Renting a plot, albeit quite different from their former farming experience, is one solution. This social context defines plot renting as a Chinese alternative food initiative, but one that is quite distinct from Western types of AFNs.

Conditions shaping China's alternative food networks

Our case studies revealed that AFNs in China demonstrated the elements of alternativeness to uneven extents. It is also critical to note that China's unique context shapes their development and alternativeness in complex ways. Within the Chinese political economy, there is "an apparently restrictive political environment in which rapid socio-economic and cultural changes are taking place" (Ho and Edmonds 2007: 2). Many confrontational and transformative strategies embedded within AFNs are adapted. Similar to the cases of environmentalism characterized by Ho and Edmonds (2007), AFNs in China display a "fragmentary, highly localized, and non-confrontational form" (p. 14). Farmers' markets, buying clubs and NGOs are moving cautiously to, in Ho and Edmonds' words, "evade even the slightest hint at organized opposition against the central Party-state" (p. 3). Hence, the political alternativeness noted by Whatmore, Stassart, and Renting (2003) is not always apparent in the Chinese context. Chinese AFNs, situated within a particular social, political, and economic background, exhibit a very different landscape of alternativeness, as we have shown in the previous section. This context that characterizes AFNs in China is shaped by three key factors.

First, in Chinese government and research circles, there is a narrow understanding of organic farming and a strong "technological managerialism" (Goodman and Goodman 2008), linked to the broader scientism and its manifestations in governmental policies. Consumers often consider organic farming merely as a farming practice that provides safe, quality food. There is a widespread concern among Chinese governmental officials and researchers that if too widely adopted, organic agriculture could jeopardize national food security by reducing productivity (Scott et al. 2014). Government policies to support the development of organic agriculture are mainly limited to infrastructural aspects (e.g. subsidies for construction of greenhouses) to promote the scaling-up of organic farms rather than improving agronomic production practices. The ecological consequences (use of plastics in greenhouses and use of energy for heating) and social consequences (exclusion of small-scale producers) of scaling up organic farms are not considered.[16] The indifference toward ecological implications also exists among many organic consumers. Our interviews with CSA farmers in Beijing and Fuzhou (Fujian province) revealed that even CSA shareholders might not develop values of "ethical consumerism." For example, a CSA farm in Fuzhou found it very hard to carry out an "organic food waste collecting" project among their shareholders due to lack of environmental awareness.[17] Although some attempts by food activists to politicize food consumption (Wilkinson 2010) could also be found in China in the form of educating consumers about their "right to know" and promoting the purchasing of organic and local food as means of "voting with your chopsticks," it was usually criticized by opponents as promoting "idealistic and unrealistic" values to the public (Sun 2013). Maintaining a non-confrontational manner is a key priority for many AFN initiatives.

The second element that characterizes the landscape of AFNs in China is that food localization—in terms of a strong concern for the provenance of food—has not yet been widely embraced among ordinary consumers in China, despite being promoted by CSAs, farmers' markets, and buying clubs. China's food system used to be very regionally oriented before the mass supermarketization process began in the 1990s (see Reardon, Berdegué, and Timmer 2005). Many Chinese have recent memories of eating seasonal food—which, in winter in northeast China, meant only cabbage, daikon radish, and potatoes. However, these conventions of food consumption have faded away in the last two decades. Being able to eat food from around the world at any time of the year is one of the many privileges of residents in large urban centres (see Garnett and Wilkes 2014). As many CSA farmers acknowledged, shareholders' main complaints have been about the limited choice of produce. It has posed a key challenge for food activists in China, despite the efforts of CSA farms to promote the "alternative" practice of eating local and seasonal food. The alternative conceptualization of "local" and "seasonal" in the West, where AFNs are well developed, is being integrated into the discourse of Chinese AFNs, but this is bound to be a long and difficult process.

The third aspect of context in which Chinese AFNs have evolved is the lack of attention to social justice concerns. Although farmers' markets and buying clubs organizers in China have an awareness of social justice in terms of opening up opportunities for farmers, consumers who are driving the development of AFNs show little interest or awareness of this priority. Many of the "new farmers" who founded the CSAs, the housewives operating the buying clubs, the organizers who run the farmers' markets, and even the urbanites who rent the plots for farming, are well-educated elites. The inclusion of "real" peasants in the construction of AFNs in China is minimal, although there are a few exceptions. The central connotations of "reconnection" implied by the current AFN literature are more a romanticization than a reflection of actual ethical values within AFNs in China. Many buying clubs and farmers' markets are merely direct procurement channels for many consumers. In many AFNs, trust is not strong between producers and consumers, and sometimes not even among producers. For example, our observations of online discussions revealed that some producers frequently accused others of cheating in ecological farming practices.

Despite this lack of trust and social justice, we have seen a strong set of core values among the small number of AFN organizers. Therefore, there is a disconnect in values between the organizers of these AFNs and their customers. This disconnect is largely due to the fact that most AFNs in China were introduced from the West, rather than being endogenous initiatives with a broad social base. The lack of strong civil society organizations in China is also a contributing factor. This is consistent with our characterization of AFNs as "consumer-driven" since the introduction of these initiatives to China was driven by consumer demands for safe food. The Western origin of these initiatives renders the "alternativeness" of them limited to date. On the one hand, the organizers who started these initiatives have to cope with the food safety concerns of consumers by demonstrating that their food is safe and healthy; on the other hand, they are also trying to help their customers to appreciate the multiple values that AFNs uphold.

The vigorous efforts of AFN food activists include striving to increase communications between producers and consumers in farmers' markets (orally or in written flyers or online), organizing "talks" held after the farmers' markets, "family experience" opportunities on CSA farms, and educational activities among buying club members. Although very nascent and limited in scope, these endeavours enable environmental and social relations to be gradually woven into consumers' perceptions of food "quality," which will lead to higher demand for quality food. In sum, the alternativeness of these nascent AFNs is evolving rapidly amid the dynamic interactions between the organizers (food activists) and customers. The landscape of alternativeness in Chinese AFNs will continue to be fluid as these networks develop and consolidate.

Discussion and conclusions

This study provides the first systematic characterization of AFNs in China, thereby providing a counterbalance to the current AFN literature that deals mainly with cases in the West (see Table 5.1). It shows that it might be oversimplified to criticize an alternative food initiative for not being alternative in terms of one or more dimensions. Rather, a closer scrutiny of more specific elements is needed. The characterization of AFNs in this study offers a framework, though it might not necessarily represent every dimension of alternativeness. This framework will be especially relevant for examining nascent AFNs in emerging economies given that much of their alternativeness is still in the early stage of formation.

Our analysis has revealed both similarities and differences between AFNs in China and the West. Chinese AFNs were found to resemble their counterparts in the West in two respects. First, like AFNs in the West, elitism is also evident in Chinese AFNs, although with different connotations. CSA operators and customers in China exhibit a strong middle-class feature. Many CSA operators are well-educated people from urban backgrounds. Second, like the existing literature, our analysis of the embeddedness of Chinese AFNs also underscores the importance of the social, political, and economic context in shaping their practices, such as the case of recreational garden plot rentals.

As for the differences between AFNs in China and the West, our analysis underscored three points. First, rather than being rooted in a fertile civil society context that has a rich discourse focused on issues of empowerment and community building (Schumilas et al. 2012; Schumilas 2014), AFNs in China emerged within the context of widespread food safety scares (see also Krul and Ho 2017). In the process of responding to consumer demands, food producers played a limited role in the emergence of AFNs in China. This "consumer-driven" feature leads to the second difference. Our unpacking of alternativeness reveals that healthfulness of food, in terms of avoiding residues and being more nutritious, is the most important element of alternativeness that propels consumers' participation in AFNs. In contrast to AFNs in the West, other elements of alternativeness associated with AFNs were not strongly evident. In particular, AFNs in China have not typically been established to oppose the globalized industrial food system. AFN customers' primary interest in the "healthfulness of food," among other elements of alternativeness, conveys "weaker alternative systems," in Watts et al.'s (2005: 30) words. Thus, Chinese AFNs face genuine threats of "incorporation and subordination" within conventional food provision channels. Third, besides the different elements of "alternativeness," Chinese AFNs differ from their Western counterparts in other ways. For example, with stronger interventions of the state, farmers' markets in China face legitimacy challenges (as we discuss in Chapter 7). Peasant farmers have also been marginalized in decision-making in CSA operations.

This chapter also identified a potential value inconsistency between the organizers of alternative food initiatives and their customers. Although the founders of CSA farms and farmers' markets have a strong desire to promote ecological, social justice, and/or political values to their customers, they understand that participation of customers in these venues is mainly driven by food safety and health concerns. Therefore, food activists in China are trying to cater to consumer concerns while also promoting a wider set of values. This inconsistency renders it difficult to form a strong solidarity between these two groups and impacts community building within these venues. However, it also opens potential space for deeper interactions between these activists and their customers.

The "consumer-driven" feature also shapes the alternativeness significantly by pitching the core attributes of alternative food initiatives as meeting food safety requirements while ecological and social values are given less prominence. Therefore, the social-political transformative potential of AFNs in China is constrained. On the contrary, what consumers are interested in matters the most. This also makes the further unpacking of alternativeness necessary given that the four major dimensions of alternativeness do not directly address specific consumer interests in food. However, this does not necessarily mean that there is no representation of ecological and social values among consumers. CSA participants also demonstrate a certain level of ecological awareness. Urbanites who rent garden plots do have a strong inclination toward reconnecting with the land and with others. But these values are weaker compared to the interest in healthfulness of food within these initiatives.

Despite the limited alternativeness in Chinese AFNs, cyberspace—especially Weibo (Chinese for Twitter-like microblogs)—is an emerging realm that enhances producer–consumer connections. Educational lectures about sustainable food behaviours are publicized online. Mobile apps, particularly WeChat (a popular Chinese app that facilitates instant communication and integrates many other services), are also extensively deployed for marketing and public education. Chinese food activists are making full use of the internet and social networks to spread information about the ecological and social alternativeness of CSAs and farmers' markets among their followers. Personal and social connections that embody "trust" are gradually permeating the landscape of AFNs in China.

Being introduced from a Western context rather than being endogenous initiatives, AFNs in China, especially CSAs and farmers' markets, are experiencing a complex process of adaptation. This process, constantly shaped by multiple stakeholders as well as economic, environmental, political, and cultural conditions, is reflected in the contested discourses or the problematization of alternative values within these AFNs. The uneven alternativeness that we analyzed is a result of this adaptation. Nevertheless, debates are ongoing, and the power dynamics within this adaptation are changing rapidly. How Chinese AFNs will evolve in the coming years is yet to be unveiled.

Notes

1 Interview with the founder of a CSA farm, December 6, 2012, Beijing.
2 In late 2012, Wen facilitated the establishment of China Rural Reconstruction Institute at Southwest University in Chongqing. CSAs and ecological agriculture have been key instruments for the so-called New Rural Reconstruction Movement. See Chapter 8 for more discussion on the New Rural Reconstruction Movement promoted by intellectuals from China Renmin University and its roles in the development of alternative food networks in China.

3 Interview with a CSA farmer, December 6, 2012, Beijing.
4 Interview with a CSA farmer and farm workers, April 1, 2012, Beijing.
5 Interview with a CSA farmer from Chongming Island, May 27, 2012, Shanghai.
6 Interview with a Beijing Farmers' Market organizer, April 3, 2012, and December 6, 2012, Beijing. We identified about 20 organic or ecological farmers' markets across the country. The frequency, popularity, and reputation of these markets differ considerably.
7 Interview with one of the Beijing Farmers' Market organizers, April 3, 2012, Beijing.
8 Interview with a CSA farmer, December 6, 2012, Beijing.
9 Interview with the founder of a buying club in Beijing, April 9, 2012, Beijing.
10 Interview with a small-scale ecological farmer, June 2, 2012, Fuzhou, Fujian province.
11 See Chapter 7 for more detailed discussion of the tensions within the Beijing Farmers' Market.
12 Interview with founders of Green Heartland, April 30, 2012, Chengdu, Sichuan province.
13 Two other prominent buying clubs in China, the Green League Mums' Buying Club and the Shanghai Caituan, were founded by and comprised mostly housewives.
14 Interview with the founder of Beijing Green League Mum's Buying Club, April 9, 2012, Beijing.
15 We have also heard about this "weekend farming" phenomenon in Japan and South Korea (*Los Angeles Times* 2010; Urban Plant Project Seoul 2010).
16 See Chapter 6 for detailed discussion on the ecological dimension of alternative food networks.
17 Interview with a CSA farmer, June 2, 2012, Fuzhou, Fujian province. The farm tried to collect organic food waste from its shareholders in order to make compost but it got little response.

References

Abrahams, C 2007, "Globally useful conceptions of alternative food networks in the developing south: The case of Johannesburg's urban food supply system." In: Maye, D, Holloway, L, and Kneafsey, M (eds.), *Alternative Food Geographies: Representation and Practice*. Elsevier, Amsterdam, pp. 95–114.

Alkon, A 2008, "From value to values: Sustainable consumption at farmers markets," *Agriculture and Human Values*, vol. 25, no. 4, pp. 487–498.

Allen, P 2010, "Realizing justice in local food systems," *Cambridge Journal of Regions, Economy and Society*, vol. 3, no. 2, pp. 295–308.

Allen, P and Guthman, J 2006, "From 'old school' to 'farm-to-school': Neoliberalization from the ground up," *Agriculture and Human Values*, vol. 23, no. 4, pp. 401–415.

Allen, P, FitzSimmons, M, Goodman, M, and Warner, K 2003, "Shifting plates in the agrifood landscape: The tectonics of alternative agrifood initiatives in California," *Journal of Rural Studies*, vol. 19, no. 1, pp. 61–75.

Barton, D, Chen, Y, and Jin, A 2013, "Mapping China's middle class," *McKinsey Quarterly*, June 2013. Accessed at www.mckinsey.com/industries/retail/our-insights/mapping-chinas-middle-class.

Beckie, MA, Kennedy, EH, and Wittman, H 2012, "Scaling up alternative food networks: Farmers' markets and the role of clustering in western Canada," *Agriculture and Human Values*, vol. 29, no. 3 pp. 333–345.

Beijing Country Fair 2011, "Introduction of Beijing Country Fair," Accessed at http://blog.sina.com.cn/s/blog_725ab7d40100xqdt.html on September 2, 2013 (in Chinese).

Brown, C and Miller, S 2008, "The impacts of local markets: A review of research on farmers markets and community supported agriculture (CSA)," *American Journal of Agricultural Economics*, vol. 90, no. 5, pp. 1296–1296.

China Agriculture Information Web 2013, "'Happy Farm' model became the growth point of infrastructure agriculture in Haidian district, Beijing." Accessed at http://nc.mofcom.gov.cn/articlexw/xw/dsxw/201305/18503935_1.html on August 3, 2014 (in Chinese).

DeLind, LB 2011, "Are local food and the local food movement taking us where we want to go? Or are we hitching our wagons to the wrong stars?" *Agriculture and Human Values*, vol. 28, no. 2, pp. 273–283.

Ding, D, Liu, P, and Ravenscroft, N 2018, "The new urban agricultural geography of Shanghai," *Geoforum*, vol. 90, pp. 74–83.

DuPuis, M and Gillon, S 2009, "Alternative modes of governance: Organic as civic engagement," *Agriculture and Human Values*, vol. 26, no. 1–2, pp. 43–56.

Escher, F, Schneider, S, and Ye, J 2018, "The agrifood question and rural development dynamics in Brazil and China: towards a protective 'countermovement.'" *Globalization*, 15(1): 92–113.

Feagan, R 2007, "The place of food: Mapping out the 'local' in local food systems," *Progress in Human Geography*, vol. 31, no. 1, pp. 23–42.

Feagan, R and Henderson, A 2009, "Devon Acres CSA: Local struggles in a global food system," *Agriculture and Human Values* vol. 26, no. 3, pp. 203–217.

Feenstra, G 1997, "Local food systems and sustainable communities," *American Journal of Alternative Agriculture*, vol. 12, no. 1, pp. 28–36.

Freidberg, S and Goldstein, L 2011, "Alternative food in the Global South: Reflections on a direct marketing initiative in Kenya," *Journal of Rural Studies*, vol. 27, no. 1, pp. 24–34.

Gale, HF 2011, "Building trust in food," *China Dialogue*. Accessed at www.china-dialogue.net/article/show/single/en/4207-Building-trust-in-food on November 12, 2012.

Garnett, T and Wilkes, A 2014, "Appetite for change: Social, economic and environmental transformations in China's food system" Accessed at www.fcrn.org.uk/sites/default/files/fcrn_china_mapping_study_final_pdf_2014.pdf on March 19, 2014.

Goodman, D 2003, "Editorial: The quality 'turn' and alternative food practices: Reflection and agenda," *Journal of Rural Studies*, vol. 19, no. 1, pp. 1–7.

Goodman, D 2004, "Rural Europe redux? Reflections on alternative agrofood networks and paradigm change," *Sociologia Ruralis*, vol. 44, no. 1, pp. 3–16.

Goodman, D 2009, *Place and Space in Alternative Food Networks: Connecting Production and Consumption*, Environment, Politics and Development Working Paper Series paper # 21, Department of Geography, King's College London.

Goodman, D, DuPuis EM, and Goodman, MK 2012, *Alternative Food Networks: Knowledge, Practice, and Politics*. Routledge, London and New York.

Goodman, D and Goodman, M 2008, "Alternative food networks." In: Kitchin, R and Thrift, N (eds.), *International Encyclopedia of Human Geography*. Elsevier, Oxford.

Guthman, J 2008, "'If they only knew': Color blindness and universalism in California alternative food institutions," *The Professional Geographer*, vol. 60, no. 3, pp. 387–397.

Hinrichs, C 2000, "Embeddedness and local food systems: Notes on two types of direct agricultural market," *Journal of Rural Studies*, vol. 16, no. 3, pp. 295–303.

Hinrichs, C 2003, "The practice and politics of food system localization," *Journal of Rural Studies*, vol. 19, no. 1, pp. 33–45.

Hinrichs, C and Allen, P 2008, "Selective patronage and social justice: Local food consumer campaigns in historical context," *Journal of Agricultural and Environmental Ethics*, vol. 21, no. 4, pp. 329–352.

Hinrichs, C and Kremer, KS 2002, "Social inclusion in a Midwest local food system project," *Journal of Poverty*, vol. 6, no. 1, pp. 65–90.

Ho, P and Edmonds, RL (eds.) 2007, *China's Embedded Activism: Opportunities and Constraints of a Social Movement*. Routledge, London.

Holloway, L, Cox, R, Venn, L, Kneafsey, M, Dowler, E, and Tuomainen, H 2006, "Managing sustainable farmed landscape through 'alternative' food networks: A case study from Italy," *The Geographical Journal*, vol. 172, no. 3, pp. 219–229.

Jarosz, L 2000, "Understanding agri-food networks as social relations," *Agriculture and Human Values*, vol. 17, no. 3, pp. 279–283.

Jarosz, L 2008, "The city in the country: Growing alternative food networks in metropolitan areas," *Journal of Rural Studies*, vol. 24, no. 3, pp. 231–244.

Jones, O, Kirwan, J, Morris, C, Buller, H, Dunn, H, Hopkins, A, and Whittington, F 2010, "On the alternativeness of alternative food networks: Sustainability and the co-production of social and ecological wealth." In: Fuller, D, Jones, A, and Lee, R (eds.), *Interrogating Alterity: Alternative Economic and Political Spaces*, Ashgate, Oxford, pp. 95–112.

Ju, H 2009, "Factors influencing consumer participations in community supported agriculture: A case study in Anlong village, Chengdu," master's thesis, Sichuan Agriculture University (in Chinese).

Kirwan, J 2004, "Alternative strategies in the UK agro-food system: Interrogating the alterity of farmers' markets," *Sociologia Ruralis*, vol. 44, no. 4, pp. 395–415.

Kirwan, J 2006, "The interpersonal world of direct marketing: Examining conventions of quality at UK farmers' markets," *Journal of Rural Studies*, vol. 22, no. 3, pp. 301–312.

Kirwan, J and Foster, C 2007, "Public sector food procurement in the United Kingdom: Examining the creation of an 'alternative' and localized network in Cornwall." In Goodman, D, DuPuis, EM, and Goodman, MK, *Alternative Food Networks: Representation and Practice*, Elsevier, Oxford, pp. 185–202.

Klein, J 2013, "Everyday approaches to food safety in Kunming," *The China Quarterly*, vol. 214, pp. 376–393.

Krul, K and Ho, P 2017, "Alternative approaches to food: Community supported agriculture in urban China," *Sustainability*, vol. 9, no. 5, pp. 844–860.

Lang, KB 2010, "The changing face of community-supported agriculture," *Culture and Agriculture*, vol. 32, no. 1, pp. 17–26.

Leung, A 2016, *The Chinese Consumer in 2030*. Economist Intelligence Unit. Accessed at www.eiu.com/public/topical_report.aspx?campaignid=Chineseconsumer2030 on April 13, 2018.

Little Donkey Farm 2012, "Weekend peasants' essays in the field." Accessed at www.littledonkeyfarm.com/forum.php?mod=viewthreadandtid=4186andextra=page%3D1 on November 12, 2012 (in Chinese).

Little, R, Maye, D, and Ilbery, B 2010, "Collective purchase: Moving local and organic foods beyond the niche market." *Environment and Planning A*, vol. 42, no. 8, pp. 1797–1797.

Liu, Fang 2012, "Looking for a new way of farming: Shi Yan and her 'Shared Harvest,'" *Green Leaf*. Accessed at http://blog.sina.com.cn/s/blog_55a11f8e0102e7ty. html accessed 12 November 2012 (in Chinese).

Liu, Fei 2012, "Institutional embeddedness and local food system: Based on a case study of three typical CSAs in Z city," *China Agricultural University Journal of Social Sciences*, vol. 29, no. 1, pp. 140–149 (in Chinese).

Los Angeles Times 2010, "Some Japanese take up weekend farming," December 6. Accessed at http://articles.latimes.com/2010/dec/06/world/la-fg-japan-farmers-20101206 on August 4, 2014.

Lyson, T 2004, *Civic Agriculture: Reconnecting Farm, Food, and Community*. Tufts University Press, Boston.

Marsden, T and Smith, E 2005, "Ecological entrepreneurship: Sustainable development in local communities through quality food production and local branding," *Geoforum*, vol. 36, no. 4, pp. 440–451.

Maye, D, Kneafsey, M and Holloway, L 2007, "Introducing alternative food geographies." In: Maye, D, Holloway, L, and Kneafsey, M (eds.), *Alternative Food Geographies: Representation and Practice*. Elsevier, Oxford, pp. 1–20.

Merrifield, C and Shi, Y n.d., CSA in China: An Introduction. *URGENCI*. Accessed at http://urgenci.net/csa-in-china-an-introduction-by-caroline-merrifield-and-shi-yan.

Murdoch, J, Marsden, T, and Banks, J 2000, "Quality, nature, and embeddedness: Some theoretical considerations in the context of the food sector," *Economic Geography*, vol. 76, no. 2, pp. 107–125.

Nelson, E, Gómez Tovar, L, Schwentesius Rindermann, R, and Gómez Cruz, MÁ 2010, "Participatory organic certification in Mexico: An alternative approach to maintaining the integrity of the organic label," *Agriculture and Human Values*, vol. 27, no. 2, pp. 227–237.

Pan, J and Du, J 2011a, "Alternative responses to 'the modern dream': The sources and contradictions of rural reconstruction in China," *Inter-Asia Cultural Studies*, vol. 12, no. 3, pp. 454–464.

Pan, J and Du, J 2011b, "The social economy of new rural reconstruction," *China Journal of Social Work*, vol. 4, no. 3, pp. 271–282.

Paüla, V and McKenzie, FH 2013, "Peri-urban farmland conservation and development of alternative food networks: Insights from a case-study area in metropolitan Barcelona (Catalonia, Spain)," *Land Use Policy*, vol. 30, no. 1, pp. 94–105.

Pei, X, Tandon, A, Alldrick, A and Giorgi, L 2011, "The China melamine milk scandal and its implications for food safety regulation," *Food Policy*, vol. 36, no. 3, pp. 412–420.

Qiao, Y 2010, "Organic farming research in China," *Organic Research Center Alliance*. Accessed at www.orca-research.org/orca-china.html on September 13, 2013.

Raynolds, LT 2000, "Reembedding global agriculture: The international organic and fair trade movements," *Agriculture and Human Values*, vol. 17, pp. 297–309.

Reardon, T, Berdegué, J, and Timmer, CP 2005, "Supermarketization of the 'emerging markets' of the Pacific Rim: Development and trade implications," *Journal of Food Distribution Research*, vol. 36, no. 1, pp. 3–12.

Rocha, C and Lessa, I 2009, "Urban governance for food security: The alternative food system in Belo Horizonte, Brazil," *International Planning Studies*, vol. 14, no. 4, pp. 389–400.

Sanders, R 2000, *Prospects for Sustainable Development in the Chinese Countryside: The Political Economy of Chinese Ecological Agriculture*, Ashgate, Brookfield, VT.

Sanders, R 2006, "A market road to sustainable agriculture? Ecological agriculture, green food and organic agriculture," *Development and Change*, vol. 37, no. 1, pp. 201–226.

Schumilas, T 2014, "Alternative food networks with Chinese characteristics," PhD thesis, University of Waterloo.

Schumilas, T, Scott, S, Si, Z, and Fuller, T 2012, *CSAs in Canada and China: Innovation and Paradox*. Presented at the International Conference on Rural Reconstruction and Food Sovereignty, 2nd South-South Forum on Sustainability, Chongqing, China, December 7–11.

Scialabba, NE and Müller-Lindenlauf, M 2010, "Organic agriculture and climate change," *Renewable Agriculture and Food System*, vol. 25, no. 2, pp. 158–169.

Scott, S, Si, Z, Schumilas, T, and Chen, A 2014, "Contradictions in state- and civil society-driven developments in China's ecological agriculture sector," *Food Policy*, vol. 45, no. 2, pp. 158–166.

Sheng, J, Shen, L, Qiao, Y, Yu, M, and Fan, B 2009, "Market trends and accreditation systems for organic food in China," *Trends in Food Science and Technology*, vol. 20, no. 9, pp. 396–401.

Shi, T 2002, "Ecological agriculture in China: Bridging the gap between rhetoric and practice of sustainability," *Ecological Economics*, vol. 42, no. 3, pp. 359–368.

Shi, Y, Cheng, C, Lei, P, Wen, T, and Merrifield, C 2011a, "Safe food, green food, good food: Chinese community supported agriculture and the rising middle class," *International Journal of Agricultural Sustainability*, vol. 9, no. 4, pp. 551–558.

Shi, Y, Cheng, C, Lei, P, Zhu, Y, Jia, Y and Wen, T 2011b, "Correlation analysis of ecological urban agriculture development and the rise of urban middle class: A participatory study based on the operation of Little Donkey Farm CSA," *Guizhou Social Sciences*, vol. 254, no. 2, pp. 55–60 (in Chinese).

Shi, T and Gill, R 2005, "Developing effective policies for the sustainable development of ecological agriculture in China: The case study of Jinshan county with a systems dynamics model," *Ecological Economics*, vol. 53, no. 2 pp. 223–246.

Shu, Q 2012, "Beijing farmers' market: an attempt to build the 'food community,'" *Wall Street Journal* (Chinese version). Accessed at http://cn.wsj.com/gb/2012 1029/TRV082620.asp on November 12, 2012.

Smithers, J, Lamarche, J, and Joseph, A 2008, "Unpacking the terms of engagement with local food at the farmers' market: Insights from Ontario," *Journal of Rural Studies*, vol. 24, no. 3, pp. 337–337.

Sonnino, R and Marsden, T 2006, "Beyond the divide: Rethinking relationships between alternative and conventional food networks in Europe," *Journal of Economic Geography*, vol. 6, no. 2, pp. 181–199.

Sun, D 2013, January 2, Microblog posts. Accessed at www.weibo.com/p/1005051071561494/weibo?is_search=0andvisible=0andis_tag=0andprofile_ftype=1andpage=116#feedtop on December 5, 2013 (in Chinese).

Thiers, P 2002, "From grassroots movement to state-coordinated market strategy: The transformation of organic agriculture in China," *Environment and Planning C: Government and Policy*, vol. 20, no. 3, pp. 357–373.

Thiers, P 2005, "Using global organic markets to pay for ecologically based agricultural development in China," *Agriculture and Human Values*, vol. 22, no. 1, pp. 3–15.

Tregear, A 2011, "Progressing knowledge in alternative and local food networks: Critical reflections and a research agenda," *Journal of Rural Studies*, vol. 27, no. 4, pp. 419–430.

Urban Plant Project Seoul 2010, September 21, "Weekend farm." Accessed at http://urbanplantseoul.wordpress.com/2010/09/21/%EC%A3%BC%EB%A7%90%EB%86%8D%EC%9E%A5-weekend-farm on August 4, 2014.

Wang, R, Si, Z, Ng, C, and Scott, S 2015, "The transformation of trust in China's alternative food networks: disruption, reconstruction, and development," *Ecology and Society*, vol. 20, no. 2, pp. 1–19.

Watts, DCH, Ilbery, B, and Maye, D 2005, "Making reconnections in agro-food geography: Alternative systems of food provision," *Progress in Human Geography*, vol. 29, no. 1, pp. 22–40.

Wen, T, Lau, K, Cheng, C, He, H, and Qiu J 2012, "Ecological civilization, indigenous culture, and rural reconstruction in China," *Monthly Review*, vol. 63, no. 9, pp. 29–44.

Whatmore, S, Stassart, P, and Renting, H 2003, "Guest editorial: What's alternative about alternative food networks?" *Environment and Planning A*, vol. 35, no. 3, pp. 389–391.

Wilkinson, J 2010, "Recognition and redistribution in the renegotiation of rural space: The dynamics of aesthetic and ethical critiques." In Goodman, M, Goodman, D, and Redclift M (eds.), *Consuming Space: Placing Consumption in Perspective*. Ashgate, Aldershot, pp. 97–120.

Wiskerke, JSC 2009, "On places lost and places regained: Reflections on the alternative food geography and sustainable regional development," *International Planning Studies*, vol. 14, no. 4, pp. 369–387.

Yang, G 2013, "Contesting food safety in the Chinese media: Between hegemony and counter-hegemony," *The China Quarterly*, vol. 214, pp. 337–355.

Ye, X, Wang, Z, and Li, Q 2002, "The ecological agricultural movement in modern China," *Agriculture, Ecosystems and Environment*, vol. 92, no. 2–3, pp. 261–281.

Yin, C and Zhou, W 2012, "We're all farmers now," *China Dialogue*. Accessed at www.chinadialogue.net/article/show/single/en/4944-We-re-all-farmers-now on November 12, 2012.

Yuan, Y 2011, "China's tycoons go farming," *China Dialogue*. Accessed at www.chinadialogue.net/article/show/single/en/4622-China-s-tycoons-gofarming on November 12, 2012.

Zhang, W 2013, "The Chinese reality of organic farmers' markets," *New Business Weekly*. Accessed at www.yogeev.com/article/29783.html on March 13, 2014 (in Chinese).

6 Economic, ecological, and interpersonal dimensions of alternative food networks

Theresa Schumilas

Contradictions and paradoxes: the puzzle of alternative food networks in China

Almost exclusively, AFNs have been understood as a Western phenomenon, where they have emerged in liberal capitalist democracies with industrialized food systems characterized by private land ownership, a declining small farm sector, consolidated farm to retail chains, predominance of supermarket retail, standards and laws ostensibly to safeguard food safety, and an extensive civil society sector organizing and advocating for changes in various ways. The Chinese context sits in contrast with its unique version of "capitalism with socialist characteristics," a commons approach to land ownership, predominance of smallholder agriculture and traditional marketing chains based on wholesale and wet markets, a focus on agricultural productivity to support an obsession with food security and increasing meat consumption, nascent food safety legislation, and a civil society with limited autonomy from an authoritarian state that keeps shifting the terrain of what is permitted. In this chapter, we suggest that much could be gained from exploring the paradoxical emergence of AFNs in this contradictory and shifting landscape.

How can we explain the recent development of AFNs in China, in the context of a state-driven, yet market-oriented economy with limited civil society involvement, amid unprecedented cultural change? In this chapter we examine the economic, ecological, and interpersonal dimensions of these AFNs, drawing on site visits, interviews, and surveys from 15 CSA farms, two farmers' markets and two buying clubs, primarily located in the more industrialized and populated areas of eastern China, along the Beijing–Shanghai corridor. Throughout this analysis, we explore how China's shifting economic, environmental, cultural, and political context influences and shapes the hybrid forms these AFNs take.

Motivations for AFNs—who are these protagonists?

The initiators and organizers, as well as the members or buyers in the CSAs, farmers' markets, and buying clubs in these AFNs, are primarily a group we could refer to as "middle-class." However, in doing so, we acknowledge this is a highly debated and contested status in China, with contrasting views on its composition, characteristics, identities and political views (Li 2010). Yet, many of the individuals we interviewed used this term to describe themselves and their members or buyers. AFN initiators are generally young people, born after 1980, and therefore raised after the "reform and opening" to the West. So, they never experienced famines, collectivized farms, food rationing, or rural hardship of the Mao era, as their parents likely did. The AFN organizers we interviewed tended to be university-educated and connected to the world through the internet and often extensive personal networks. They refer to themselves as "new farmers"; this term was adopted in the title of the Third National CSA Symposium in 2011 and is frequently used in news reports describing this alternative movement.

One of the most salient aspects of the CSAs, buying clubs and farmers' markets that comprise China's AFN movement is that the motivations and ethics of the *organizers* can be contrasted with those of the *members or consumers* with whom they are trying to forge connections. The initiators of China's AFNs are driven by diverse motivations. A desire to support their livelihood intermingles with more egalitarian motives. They are concerned about the marginalization of peasants in rural China and seek to reconnect with the rural by rekindling lost food and farming traditions, and to reconnect urban consumers with land and food production. Primarily urban-born, they have limited direct experience with China's traditional peasantry, yet they feel sympathetic to its problems and see food initiatives as a way to assist. Second, AFN initiators are concerned about environmental issues and some work in collaborative relations with environmental NGOs. They seem strongly motivated by the traditional Chinese pastoral and idyllic imaginaries and lament the loss of traditions and food skills that is accompanying the modernization of the food system. Third, they are concerned about the safety of the food supply and see this as a growing "crisis" in China, and a primary way to engage with and broaden the awareness of others.

In comparison, however, as shown, members and buyers who engage with China's AFNs most likely share only the concern for food safety with the AFN initiators. Most are not particularly motivated by environmental concerns. Nor are they necessarily seeking relationships with producers. Nor are they motivated to improve the plight of peasant farmers. Indeed, I found that many AFN participants distrust China's peasantry, and it is an ongoing challenge for AFN organizers to engage these consumers in broader food justice and environmental issues. Yet, AFN organizers know that their own motives are different from those of many of the other participants in these networks, and, as will be detailed, they continually act to draw others into deeper connection in the networks.

A unique feature of the CSA farms in these networks, and one of the ways that "self-interested" CSA members are drawn into deeper connections, is through the practice of "weekend farming." Almost all of the CSAs we visited, for example, embraced a type of agritourism in which they rented plots on the farm to their members who wished to grow their own food. The table illustrates the motivations of this group of participants, as with the motivations of AFN organizers, extend beyond merely food safety, although they do not have the same broad egalitarian motivations as most of the AFN initiators. For this reason, this typology suggests that there are multiple organizer/initiator identities as well as multiple producer/consumer identities becoming entangled in these networks. These complexities in China's AFNs can be unpacked by considering their economic, ecological, interpersonal, and political relations.

Diverse economic relations

China's AFNs demonstrate diverse economic relations, shaped by the commons approach to land and the dual urban–rural citizenship (*hukou*) of agrarian reform processes described in Chapter 2. AFNs are emerging in a context where processes of "de-peasantization" and agrarian capitalism are beginning, and scholars have argued that market-based alternatives like AFNs are inevitably co-opted or mainstreamed by these processes. However, following emerging thinking on post-capitalist diverse economies (Gibson-Graham 2008), we recognize the diversity of economic life and argue that these emerging AFNs demonstrate hybrid relations, where capitalist and other-than-capitalist forms are entangled and that present opportunities to reproduce the peasantry rather than eliminate it. Indeed, there are several characteristics that distinguish Chinese AFNs from mainstream economic relations. This diversity is revealed in three ways: first, through the ways in which land and labour are treated as common pool resources; second, through the focus on livelihood, self-financing, and decentering of surplus; and, third, through a prevalent discourse on the social economy.

Land–labour nexus

Since land remains a common pool resource in China and the HRS places decisions about rural land use squarely with rural villages, urban entrepreneurs seeking to start CSAs are faced with negotiating land tenure agreements with peasants, resulting in a more complicated set of power relations than seen in waged labour relations in the West. In China, landless entrepreneurs need to seek permission to use land from peasants. In the context of China's rapid urbanization, where peasants with land rights

Figure 6.1 A local farmer hired at a CSA farm in Beijing.

in peri-urban areas are often waiting for lucrative compensation due to them when the state expropriates land for development, these leases are getting tougher and tougher to negotiate. Thus, land is a bargaining chip for peasants in China, in contrast to workers in capitalist class relations, who do not have direct control over the means of their own production and need to sell labour for wages. Several urban CSA operators explained how, on the outskirts of a large and growing city like Beijing, land leases are becoming more expensive and shorter in duration. It is also interesting that, unlike in the West, where land and labour are usually separated (in that both or either are available for purchase), land and labour in China can be a "package deal" for entrepreneurs starting CSAs. Sometimes, when villages are approached by an urban entrepreneur about leasing land, the terms demanded by the village include employment on the farm for a certain number of peasants.

This commons approach to rural land complicates the economic relations. Land is not simply a means of production or a cost for the entrepreneur as understood from a capitalist perspective. Rather, land is seen as a social safety net by the villagers who hold usage rights. Scholarship in the West typically understands land access as an elite attribute and much AFN scholarship sees injustices in ways in which women, people of colour, or the poor are excluded in AFNs because of their lack of access to land (Bedore 2010). In China's AFNs a more complicated dynamic emerges. While peasants remain marginalized in many ways, the demand for land to start CSAs by urban entrepreneurs begins to change these power relations. The peasant farmers and villagers in these AFNs are not dispossessed villagers at the service of urban food projects, as one might think of migrant labourers in the West, for example. Rather, the commons approach to land gives at least a small degree of control to marginalized villagers and peasants, while "privileged" urban entrepreneurs, seeking to respond to market opportunity, need to negotiate for it. In this way, these land relations are a reversal from those studied in Western AFNs, where a "moral and economic primacy over farming and other occupations" characterizes "American agrarianism" and results in inequities of private land ownership (Allen 2010: 300).

Focus on livelihoods and decentering surplus

As scholars have documented in the West (DeLind 2003; Feagan and Henderson 2009; Galt 2013), CSAs typically are not engaging in the same profit-maximizing logic that characterizes their mainstream counterparts. In China as well, AFN farms are focused on building sustainable livelihoods. Alternative economic scholars suggest that the degree to which alternative capitalist and non-capitalist practices and institutions can sustain livelihoods is a key measure of their economic significance (Fickey 2011). I found a strong focus on livelihoods among the CSA farmers we interviewed. Further, all of the larger (over 50 shares) CSA operators responding to a survey we conducted indicated that their CSA operation was their primary source of income, contributing over 75 per cent of the revenue in their households.

This focus on livelihood or "making a living" guards against the capitalist practice of surplus being accumulated and removed from the community. These AFNs effectively limit the flow of surplus out of the network by distributing wages to the villagers as described above, and by relying on self-financing. All but one of the CSAs, buying clubs, and markets we visited were self-financed. While some of the farms received state support for infrastructure enhancements on their farms (greenhouse construction in particular), there were no private investors to influence or extract surplus from these networks. This leaves greater surplus for reinvestment in the networks and indeed reinvestments into the farms were extensive. AFNs were using surplus revenue to invest in new cropping approaches, learn new ecological farming methods and practices, buy books and resource material, organize training events, purchase food to distribute to families in need, and/or hire more villagers as workers. This reinvestment of surplus into social and ecological improvements, or "growth by deepening" (van der Ploeg, Ye, and Schneider 2012), versus expansionary growth, has been noted in Western CSA research as well (Cameron 2010). It represents a "reservoir of social wealth" (Gibson-Graham 2001: 26) that opens up possibilities in these networks. Further, it keeps these networks autonomous and only "partially integrated" with dominant capitalist economic forces (Zhang and Donaldson 2010).

Discursive construction of the "social economy"

While China's AFNs have a strong pragmatic emphasis, they cannot be characterized *only* as such. In addition to the material practices in these networks described above, we found a pervasive discourse on the "social economy" throughout our interviews, illustrating the ways in which China's AFNs are trying to negotiate what they perceive as a contradiction between market-oriented projects and social goals. CSA operators, farmers' market volunteers, and buying club organizers in these networks all relayed a tension between market pragmatics and the ideals of a new movement they are trying to build. Interviewees struggled to find language that best describes their networks, often using the phrase the "social economy" in trying to describe a space between capitalist and state-socialist. Indeed, academics associated with these networks (Hale 2013; Pan and Du 2011) as well as academics from the West (Amin 2009; Quarter 2010) employ the social economy as a construct to describe this alternative space and initiatives that engage in market-based activities as a means of addressing community needs. In China, however, there is no legal framework that legitimizes this space, so it exists only in people's ideas.

Searching for a "social economy" is also evidenced by the way in which some operators in these networks struggle with consumerist ethics in their CSAs. Registering as NGOs is "almost impossible," so they need to rely on market-based exchanges to earn operating funds. However, the same middle-class consumerist ethics that make these networks possible are not easily accepted by many AFN participants, who are frustrated by what one person described as "consumer domination." This is a conundrum the AFN organizers and producers struggle with as they try to evoke a different ethic. To illustrate, the guidelines for one CSA explicitly state, "We do not 'regard consumers as god.' Each one of us is a part of this social movement. We and our members are not simply selling-purchasing agents, we are equal partners, and we trust each other." Such comments coupled with grasping for a social economy illustrates the "other-than-capitalist" ethics in these networks.

This analysis illustrates that AFNs in China are characterized by economic diversity, wherein capitalist relations involving waged labour, financial investment, and surplus extraction co-exist in exchange relations with peasant economies characterized by self-labour, self-provisioning, a focus on livelihoods, and spaces of peasant empowerment. These AFNs are built on a foundation of a commons approach to land. In peri-urban China, where land is in high demand, landless, urban entrepreneurs seeking to capitalize on what they perceive to be a direct-to-consumer marketing trend are placed in a position of negotiating with peasant villages to access land and labour. The results are mixed. While CSA operators lament the rising cost of land and the difficulties in negotiating for it, many of these entrepreneurs are actively expanding their farms, suggesting that access to land is not a barrier. At least in a small way, having to negotiate with peasants for land use serves to moderate urban–rural power relations. Further, while capitalist commodity relations are evident in these networks, and are perpetuated by consumer subjectivities and concerns over price, convenience, and product quality, we do not see the same path toward de-peasantization that some argue is occurring more broadly in China's agrarian economy.

Ecological relations—modern and traditional

This section looks more closely at the ecological relations in China's AFNs, with particular attention to the ways in which they define and manifest organic production approaches in the context of a state-led, technologically driven productivist approach to ecological agriculture development. First, drawing on expert consensus on key indicators of functional integrity in agricultural systems, we look at the ways in which farming systems in these AFNs enhance biological diversity, demonstrate closed-loop systems, and protect soil and water resources (Luttikholt 2007). Second, we examine the ways in which lay participants in these networks conceptualize and negotiate what they understand as organic and ecological. Here, we detail the ways in which AFN participants are contesting the state-led organic regulatory process and using civic process to codify practices and define organic for themselves. I suggest that practices in these AFNs, while being strongly ecological based on the functional integrity indicators, are nonetheless influenced by the dominant focus on productivism and are missing some fundamental ecological practices as a result. At the same time, there is an extensive adoption of traditional practices that the state-endorsed version of ecological agriculture has abandoned. Further, reacting to a widespread distrust of state-led organic and ecological agriculture intuitions, producers and consumers alike in these networks are forging bottom-up alternatives to ensure transparency, reconfigure state standards, and construct their own meaning of "organic."

Ecological with productivist pressures

Most practices on these farms appear largely consistent with organic principles and the Chinese organic standard including the practices of sourcing non-GMO seed, using on-farm composting and manure, mixed cropping, and avoidance of synthetic pesticides and fertilizers. However, the use of cover crops and intercropping, two practices considered strongly ecological and essential to organic systems, seem to be lacking in these networks. The absence of practices to protect soil resources stands out as problematic and illustrates a logic of intensification that contrasts with other traditional practices observed. The CSA operators we interviewed felt that the practices of cover cropping and intercropping are very labour demanding, and

Figure 6.2 Farmland with no cover crop in winter in a CSA farm in Beijing.

they would need to spend more time in the field in order to manage the increased complexity of these cropping systems. Further they explained that they lack equipment such as tractors necessary for these practices and the traditional knowledge of using work animals with plows has been lost. This pragmatism echoes other research on ecological agriculture in China. A cross-country case study involving China and Brazil also found Chinese cropping systems to be strongly influenced by market pressures, and thus falling short of organic principles established by IFOAM (Oelofse et al. 2011). Western scholars have considered such examples a weakening of organic practices and an unravelling of the organic movement philosophy by pragmatists responding to market pressures. In China's AFNs, this pressure to produce is not only economic. While the dominant practices in these networks are strongly ecological, the absence of some key practices suggests that the productivist ideology of the state permeates the largely traditional practices in these networks. As one farmer noted when we asked about the absence of cover crops,

> On a small farm, it is just inefficient to grow a crop that you don't sell. Plus I think most of these farmers know that we have a lot of people to feed in China. It is our responsibility to produce as much as we can. We don't have room for crops we don't eat.
>
> (Interview with a CSA farmer in Beijing, March 31, 2012)

Searching for traditional knowledge

Site visits to AFN farms revealed extensive use of closed-loop farming systems which operators referred to as "circular farming." Embracing closed-loop systems is an illustration of ties to traditional Chinese practices (Li, Liu, and Min 2011) and stands in contrast to industrialized farming systems, which tend to specialize and increase the use of off-farm inputs. Prior to China's industrialization of agriculture, these were the dominant practices of the peasant farming system as described by King (1911). A number of these circular farming techniques were evident on the CSA farms we visited. For example, almost all these farms were aerobically fermenting vegetable waste and using composted human manure for fertility. In addition, typically farms were integrating livestock into their farming systems and relying on plant-based medicines made on the farm to treat the animals.

Yet the low-input, traditional practices on these farms co-exist with advanced technologies. All the farms had extensive modern infrastructure, including modern greenhouse operations, paved roadways, concrete irrigation ditches, on-farm restaurants, farm stores, and sometimes accommodation for visitors. In some cases, it is easy to understand the reasons for this mixture of traditional and modern. This infrastructure supports multifunctionality on these farms and a strong orientation toward tourism. Modern toilets that separate liquid and solid waste make the traditional practice of separating "night soil" from urine (somewhat) easier. Similarly, modern distilling equipment adds sophistication and precision to the processing of traditional medicines. These examples reflect a general philosophy that traditional practices can be rearticulated with modern practices and thereby improved.

However, other traditional farming practices documented by King (1911) are notably absent on these farms. For example, CSA operators talked about challenges in sourcing non-GMO seed, but none of the CSA operators we interviewed talked about saving seed, and only one farmer was raising heritage breed animals. Clearly, seed saving and variety breeding are central to the re-establishment of traditional practices, so the absence of these practices is curious and problematic. Further, there were few traditional pest management approaches being used on these farms, even though several of the farmers told us that managing pests was their biggest challenge. But we saw none of the traditional practices that would help address their challenges. For example, there were no insectaries planted to draw in beneficial insects. Nor were there any symbiotic cropping patterns, like the use of frogs to control insects, as described by King (1911). Indeed, the cropping patterns observed were rather unimaginative and pragmatic. Vegetables were conveniently planted in rows to facilitate harvest and, as noted above, intercropping was limited.

There are some possible explanations as to why some traditional practices are adopted and others are not. First, these are nascent farms and inexperienced urban farm operators. Some traditional practices, like the complex cropping systems and plant breeding, described by King (1911), might be, for the present, beyond the skill levels of these new farmers. Second, the state strongly influences the adoption of particular practices through its subsidies for certain technologies. In particular, in recent years, the state has recognized the potential of farm-based tourism for economic development and has supported its expansion (Su 2011). This policy direction and associated funding could explain the extensive on-farm infrastructure seen on these CSAs and the adoption of related practices. However, with neither funding nor extension support, traditions such as seed saving or preserving heritage livestock breeds are not being preserved in these networks.

Resistance to the state's organic standard

Few of the CSA farms we visited chose to have their ecological processes verified by a third party and hence certified under China's organic regime. Instead of supporting the national standard, most of the CSA, farmers' market, and buying club operators we spoke with were cynical about the state's role in organic standard-setting and speculated that its development was not motivated by ecological concern. They felt the organic standard was designed to placate widespread public concern with food safety, was plagued by corruption, and was unaffordable for most farmers.

Reacting to these concerns and exclusion from standard-setting processes, producers and consumers in China's AFNs are contesting and reconfiguring state standards by constructing their own meaning of organic. Rejecting the state's expert-led third-party verification system, their approach relies on the development of lay knowledge. Both farmers' markets and buying clubs in these networks have developed practices they consider fundamental to organic production along with regimes for verifying these practices. The criteria for this verification are not well defined and codified as in a national organic standard. Instead, they convey the general interest of farmers' markets and buying clubs volunteers in supporting a shift toward a less-industrialized and safer food system. These volunteers have translated the technical aspects of the organic standards into lay language with various indicators, which they use to inspect farms. They are committed to transparency and post the criteria and the results of their farm visits on their websites.

This process of community-based standard development reflects a civic approach wherein expertise is not limited to experts with credentials but rather is a shared responsibility inclusive of lay perspectives. Citing examples of corruption, participants in the AFNs we interviewed distrust the state's standard-setting mechanism and the bureaucracy charged with its enforcement. In what can be understood as a form of everyday resistance to the state's approach, this distrust is motivating the formation of nascent civil society action to develop standards in which they can place their trust. Further, these AFN processes are situated and reflexive versus universal and inflexible, allowing for exceptions in particular situations. So, while it could be argued that such civic processes in effect "let the state off the hook" by accepting its responsibilities (Guthman 2008), it can also be argued that in this process people are developing skills that could be prefiguring future democratic processes.

In conclusion, China's AFNs articulate a mixture of traditional and modern production methods in a type of ecological hybridity. In contrast to AFNs in North America and Western Europe, China's networks have evolved in the absence of any type of ecological social movement. The state has sponsored the development of a "made in China" ecological agriculture sector. In efforts to build stronger import and export markets for high-quality foods, the state has built a complex set of standards, which are largely devoid of other policy support. The state's reach and drive for productivism extend to these "alternatives" and CSA operators and farmers eschew key ecological practices because they would negatively impact yields. In the absence of organized civil society or government support, these producers, many of whom are new to farming, are challenged in a context where traditional practices are being lost. Yet, in the face of widespread distrust of and exclusion from the state regulatory system for organic food, these face-to-face interactions have spawned new approaches to defining organic and verifying production practices. In resistance to state-authorized standards, consumers and producers in these AFNs are co-constructing what organic means and building skills and establishing trust in the process. In this way, paradoxically, the state's focus on food sufficiency and productivism, and food quality through the setting of artificially high standards, results in the formation of nascent civil society that is resisting the dominant paradigm by constructing its own understanding and practice of organic from below.

Interpersonal relations—the ethics of care

China's food safety crisis and associated "food fears" are promoting attempts to rebuild trust in food and this is illustrated by the rapid emergence of AFNs. In this section we use an ethics of care lens to illustrate the complex interpersonal relations that are evolving in China's AFNs. We argue that, given the high level of social anxiety about food in the current context, people are coming to China's emerging AFNs to avoid unsafe food and mistrust of the food supply, rather than being drawn toward trust. In engaging in these networks, most consumers are motivated by "caring for themselves" and their immediate families. Yet, regardless of what brings people to these networks initially, for some there is a deepening of interpersonal relations and development of trust that mirrors research from the West. In this way, AFNs function as a "window" through which distrusting and self-interested consumers can enter and encounter a different ethic, which for some deepens their interpersonal connections.

"Reconnections," that is, the tangible and intangible qualities of connections between and among producers, consumers, and food production through local, direct exchange (as in for example CSAs, buying clubs, farm shops, and farmers' markets), have been extensively explored and contested in existing AFN scholarship (Cox et al. 2008; Feagan and Morris 2009; Feagan and Henderson 2009; Hendrickson and Heffernan 2002; Hinrichs 2000; Kneafsey et al. 2008). AFN scholars most frequently draw on theories of embeddedness in discussing these connections and reconnections (Granovetter 1985, Polanyi 1944). More recently, however, and in response to some of the critiques of embeddedness as an analytic, the feminist theory of "ethics of care" (Gilligan 1982; Lawson 2007; Tronto 1993) has offered another tool (and indeed the lens used in this analysis) to unpack relations in food systems. Indeed, in the West, ethics of care and trust have been seen as defining characteristics of alternative food procurement networks (cf. Kneafsey et al. 2008). Care ethics makes a useful lens through which to examine interpersonal relations in Chinese AFNs because it moves well across different political philosophies (Robinson 2010). In particular, care ethics theory challenges the dichotomy between liberalism and socialism and between the individual and the collective, because it focuses on people's ability to fulfill responsibilities to others (Robinson 2010). In this respect, comparative philosophy scholars also note that much of traditional Chinese thinking overlaps with concept of ethics of care (Luo 2012).

Taking care of self and family in a context of distrust

Consumers in these networks are motivated to do the best for themselves and their bodies in an environment of distrust in food they perceive to be frequently adulterated. As discussed, China's AFNs exist in a culture of uncertainty, where the food and the food providers are seen with suspicion. In this heightened distrust in food, which we encountered in every interview and site visit, it is the avoidance of distrust, not the positive motivation of trust, that draws consumers to these networks. To illustrate, one consumer we spoke with at farmers' markets told us that buying food at the market directly from a farmer is the "least bad" option. When we asked her about trust she responded that she could not say she *trusted* the farmers at the market, but rather that she *distrusted them less* than others.

Producers and AFN organizers recognize that consumers are drawn into these AFNs because of food safety concerns and believe that this pragmatic reconnection is a precursor for their involvement in the network. Producers and farmers' market volunteers, for example, are continually trying to maintain shift consumers to a place of caring that extends beyond their own families. Indeed, a significant amount of effort goes into such persuasive communications with market shoppers. Similarly, CSA operators are constantly trying to forge connections with members, and with consumers who are not currently members, to create an ethic of care about food. For example, all the CSAs we visited offer a calendar of diverse events as well as newsletters and blog posts that go beyond simple promotions of their goods. We do not suggest that *all* CSA members engage in these "caring about" relations. Indeed, CSA operators, market managers, and buying club organizers alike explained their challenge with establishing connections with consumers given their strong suspicions of and distrust in food relations generally, making the process of deepening the care and trust relations difficult. In North America as well, many AFNs offer events and activities in the hope of drawing members into "care-ful" reconnections but such initiatives are often met with lacklustre participation (DeLind 2003; Feagan and Henderson 2009).

Taking care of self through reconnecting to land

While reconnections and ethics of care between consumers and producers seem weak in China's social context of distrust, different kinds of ethics and reconnections seem to be more strongly enacted in these networks. Interviews, site visits, and surveys illustrate that, for many consumers, participation in these networks is a way of reconnecting, not to farmers but rather to land. On the majority of the CSAs we visited, the operator sets aside a portion of the land, sometimes as much as one-third of the farm, for members and/or non-members to rent a plot and grow their own food with the assistance of the CSA farmers. Described as China's "weekend farmers," this plot renting is part of a larger countrywide obsession with *nong jia le*, a popular form of agritourism in which middle-class urban consumers visit farms for relaxation and solitude. *Nong jia le* is translated into phrases such as "happy farm family" (Sia et al. 2013) or "delights in farm guesthouses" (Park 2008) and is a state-supported, cultural rural tourism trend in China and also South Korea, Taiwan, and Japan. The weekend farmers at these CSA farms, like *nong jia le* tourism, draw on contrasts between rural and urban life. This ethics of care for land is motivated by complicated values and beliefs. For some, this reconnection to land appears as another enactment of caring for oneself and thus might be considered as a form of respite or perhaps escapism from intensely urbanized environments. Indeed, on these farms the landscape itself seems to be more of a commodity than the vegetables, wherein consumers come to the farm to enjoy open space and fresh air while removing themselves from the dirtiness of food production. However, for others it could be that this connection to land also links to personal well-being, supporting the idea that it is a demonstration of ethics of care for oneself. Indeed, Tronto (1993) understands ethics of care for one's body and one's health not as selfishness but rather as a foundation upon which the ethics of care for others and the non-human world can be built. Indeed, on many of the farms we visited, groups of members were assembled in collective exercise such as tai chi. In other examples, China's weekend farmers are motivated by producing food they can trust. Having a plot of land to grow food on a CSA farm offers a chance to learn about and develop skills related to ecological farming. So there is a pragmatic side to the more emotional values of respite and relaxation.

Considering these examples, it is interesting that, while ethics of care for producers seems largely absent in these networks, ethics of care for land seem strong and variously motivated. Through these AFNs, people are reconnecting to and caring for land for material reasons (safe food in a context of uncertainty), for symbolic reasons (source of peace and respite), and for personal health reasons. These findings also suggest an entanglement of producer–consumer identities in these networks, echoing existing research in the West (Renting, Schermer, and Rossi 2012).

It is important to note that this absence of trust is not saying that consumers in these AFNs don't care or don't have ethics. Rather, we are suggesting that, in terms of procuring food through these "alternative" networks in China, many consumers are motivated by the instrumental needs to "take care of" themselves rather than being drawn into connections with the farmers in these networks. However, what is most interesting in this context is that CSA operators, farmers' markets volunteers and buying club operators in these AFNs seem undaunted by this lack of trust and are persevering with continuous activities and ongoing internet posts through which they seek to build reconnections and care and trust in these networks. Interviewees detailed an impressive list of festivals, workshops, events, conferences, and other activities all aimed at building what seems like hard fought for reconnections.

Deepening care and co-constructing the network

While it seems that reconnections of care and trust are not guaranteed in China's AFNs, for some consumers, relations of care seem to deepen to these final stages of care relations over time. In particular, participants who are now taking organizing roles in these AFNs reflected on their own experience of care ethics and connections. They describe how they were initially drawn into alternative food procurement because, like many others, they were trying to find safe food, but they gradually became more involved in the network and in environmental issues more broadly, just as Kneafsey et al. (2008) uncovered in their research. Distrust and care about one's *own* health motivated actions to connect, which then deepened to an ethics of care with others. As women entered these AFNs as consumers, their role gradually changed and they became producers responsible for reconnecting others to food. Existing studies in the West also underscore

Figure 6.3 Family activities at Little Donkey Farm in Beijing.

this entanglement of producer–consumer identities (Renting, Schermer, and Rossi 2012).

In conclusion, our interviews and site visits revealed how relations in Chinese AFNs demonstrate pragmatic reconnections where people care about healthy and safe food and are drawn to CSAs, markets, and buying clubs. In the context of pervasive uncertainty about food quality, these are connections motivated by distrust in the dominant food system, trust is indeed hard to build. Consumers are more likely to connect to land as weekend farmers pursuing a rural idyll than to the traditional farmers who grow their food. Undaunted by the absence of trust, however, AFN volunteers and CSA operators offer a continuous menu of activities and projects to draw people deeper into relations. For some, connections in these networks progressively deepen, identities of producer and consumer become entangled, and some people establish relations of care and trust with others. In this way, these findings echo the work of Klein (2013) and Chen (2013a, 2013b) in suggesting that, in response to disconnections through China's reform processes and the ensuing food safety crisis, people are actively seeking to rebuild ties and connections. The perception of a food safety crisis is thus stimulating the formation of AFNs as nascent forms of civil society organizations focused on food.

Conclusion

This chapter has presented an early account of new forms of producer–consumer provisioning networks emerging in peri-urban China. We have further explored the "alternativeness" of these networks along economic, ecological, and interpersonal dimensions by drawing on three different theoretical perspectives—diverse economies, functional integrity, and ethics of care. The findings complicate dualisms and binary thinking, and we have argued that, instead of fitting into the "either or" categories, China's AFNs need be seen as hybrid systems with a "yes and also" nature. The unique "alternativeness" of these AFNs is shaped by a particular assembly of economic, environmental, cultural and political conditions or characteristics of China's context, as shown in Figure 6.4.

Economic changes and globalization in China's process of reform both drive and restrain the development of AFNs. Agrarian reforms, specifically the egalitarian distribution of land and the *hukou* approach to citizenship, set the stage for China's explosive economic growth in the reform period. Not only did these changes open the door to capitalist relations and the development of a middle class; they also made land available for AFN initiation. On the one hand, China's AFNs are made possible by urbanization and a middle class in pursuit of better food quality. But, on the other, their consumerist and individualist ethics present challenges for AFN development. The impacts of globalization and China's "opening" are also mixed. While these processes have presented challenges for smallholder agriculture, at the same time global connectedness makes it possible for egalitarian minded urban consumers to draw AFN models and support from outside China.

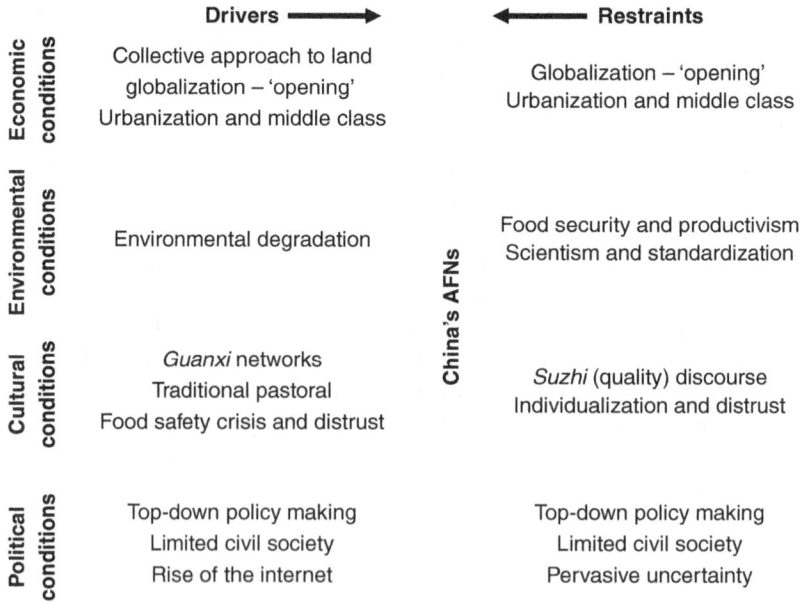

Figure 6.4 Conditions shaping China's AFNs.

China's reforms have been underpinned by an uncritical view of science and technology and a preoccupation with food security that have reinforced a productivist approach to agriculture as the foundation of the socialist market economy. Embracing productivism wholeheartedly has resulted in negative environmental impacts. China's productivist legacy is only just being revealed, and the very changes that helped production soar to meet food security goals may now be posing the barriers to meeting those goals in the future. While the resulting environmental degradation is a strong motivator for the formation of AFNs, the strong ideology about productivism, technology, and need to be food secure shapes and limits the ecological practices in China's AFNs. Further, the dream of agricultural modernization and the state's focus on food security through science and technology has driven traditional agriculture to the margins, paradoxically recalling a traditional pastoral or idyll that motivates urban producers and consumers alike.

Cultural conditions are also strong influencers of China's AFNs, often working in contradictory ways. By "cultural conditions" I mean the meanings, beliefs, ideas, and practices of everyday life. This research has identified three cultural threads, or sets of meanings, that intersect in complex ways with political economic conditions, to give China's AFNs their particular characteristics. As noted in Figure 6.4, the traditional Chinese pastoral or rural idyllic, the discourse of *suzhi* (quality), and the traditional practice of guanxi suggest a central, and often overlooked, role for such cultural tropes in explaining AFNs, in China as well as elsewhere.

Lastly, the top-down decision-making, and pervasive uncertainty of the authoritarian state has profound influences on the development of AFNs. The absence of civil society institutions, the bureaucratic requirements for NGO registration, and the seemingly excessive state oversight of mundane activities like holding farmers' markets have no doubt slowed the development of civil society–based food alternatives in China. Yet, at the same time, the absence of civil society institutions and the state's top-down decision-making are prompting democratic action in these networks such as civil society organizing, bottom-up standard development, and acts of everyday resistance.

References

Allen, P 2010, "Realizing justice in local food systems," *Cambridge Journal of Regions, Economy and Society*, vol. 3, no. 2, pp. 295–308.

Amin, A 2009, *The Social Economy: International Perspectives on Economic Solidarity*, Zed, London.

Bedore, M 2010, "Just urban food systems: A new direction for food access and urban social justice," *Geography Compass*, vol. 4, no. 9, pp. 1418–1432.

Cameron, J 2010, "Business as usual or economic innovation?: Work, markets and growth in community and social enterprises," *Third Sector Review*, vol. 16, no. 2, pp. 93–108.

Chen, W 2013a, "Perceived value of a community supported agriculture (CSA) working share: The construct and its dimensions," *Appetite*, vol. 62, pp. 37–49.

Chen, W 2013b, "Perceived value in community supported agriculture (CSA): A preliminary conceptualization, measurement, and nomological validity," *British Food Journal*, vol. 115, no. 10, pp. 1428–1453.

Cox, R, Holloway, L, Venn, L, Dowler, L, Hein, J, and Kneafsey, M 2008, "Common ground? Motivations for participation in a community-supported agriculture scheme," *Local Environment*, vol. 13, no. 3, pp. 203–203.

DeLind, L 2003, "Considerably more than vegetables, a lot less than community: The dilemma of community supported agriculture." In: Adams J (ed.), *Fighting for the Farm: Rural America Transformed*. University of Pennsylvania Press, Philadelphia, pp. 192–206.

Feagan, R and Henderson, A 2009, "Devon Acres CSA: Local struggles in a global food system," *Agriculture and Human Values*, vol. 26, no. 3, pp. 203–217.

Feagan, R and Morris, D 2009, "Consumer quest for embeddedness: A case study of the Brantford farmers' market," *International Journal of Consumer Studies*, vol. 33, no. 3, pp. 235–243.

Fickey, A 2011, " 'The focus has to be on helping people make a living': Exploring diverse economies and alternative economic spaces," *Geography Compass*, vol. 5, no. 5, pp. 237–237.

Galt, R 2013, "The moral economy is a double-edged sword: Explaining farmers' earnings and self-exploitation in community-supported agriculture," *Economic Geography*, vol. 89, no. 4, pp. 341–365.

Gibson-Graham, J 2001, *Community Economies Collective*. Accessed at www.communityeconomies.org on February 1, 2014.

Gibson-Graham, J 2008, "Diverse economies: Performative practices for 'other worlds,' " *Progress in Human Geography*, vol. 32, no. 5, pp. 613–632.

Gilligan, C 1982, *In a Different Voice: Psychological Theory and Women's Development*, Harvard University Press, Cambridge, MA.

Granovetter, M 1985, "Economic action and social structure: The problem of embeddedness," *American Journal of Sociology*, vol. 91, no. 3, pp. 481–510.

Hale, M 2013, "Reconstructing the rural: peasant organizations in a Chinese movement for alternative development," PhD thesis, University of Washington, Seattle, WA.

Hendrickson, M and Heffernan, W 2002, "Opening spaces through relocalization: Locating potential resistance in the weakness of the global food system," *Sociologia Ruralis*, vol. 42, no. 4, pp. 347–368.

Hinrichs, C 2000, "Embeddedness and local food systems: Notes on two types of direct agricultural market," *Journal of Rural Studies*, vol. 16, no. 3, pp. 295–303.

King, F 1911, *Farmers of Forty Centuries*, Rodale Press, Pennsylvania, PA.

Klein, J 2013, "Everyday approaches to food safety in Kunming," *The China Quarterly*, vol. 214, June 2013, pp. 376–393.

Kneafsey, M, Cox, R, Holloway, L, Dowler, E, Venn, L, and Tuomainen, H 2008, *Reconnecting Consumers, Producers and Food*, Berg, Oxford.

Lawson, V 2007, "Geographies of care and responsibility," *Annals of the Association of American Geographers*, vol. 97, no. 1, pp. 1–11.

Li, C 2010, "Characterizing China's middle class: Heterogeneous composition and multiple identities." In: Li C (ed.), *China's Emerging Middle Class*, Brookings Institution Press, Washington, DC, pp. 135–157.

Li, W, Liu, M and Min, Q 2011, "China's ecological agriculture: Progress and perspectives," *Journal of Resources and Ecology*, vol. 2, issue 1, pp. 1–7.

Luo, S 2012, "Setting the record straight: Confucius' notion of *Ren*," *Dao*, vol. 11, no. 1, pp. 39–52.

Luttikholt, L 2007, "Principles of organic agriculture as formulated by the International Federation of Organic Agriculture Movements," *NJAS-Wageningen Journal of Life Sciences*, vol. 54, no. 4, pp. 347–360.

Oelofse, M, Hogh-Jensen, H, Abreu, L, Almeida, G, El-Araby, A, and Hui, Q 2011, "Organic farm conventionalisation and farmer practices in China, Brazil and Egypt," *Agronomy for Sustainable Development*, vol. 31, no. 4, pp. 689–698.

Pan, J and Du, J 2011, "The social economy of new rural reconstruction," *China Journal of Social Work*, vol. 4, no. 3, pp. 271–282.

Park, C 2008, "Delights in farm guesthouses: Nongjiale tourism, rural development and the regime of leisure-pleasure in post-Mao China," PhD thesis, University of California, USA.

Polanyi, K 1944, *The Great Transformation: The Political and Economic Origins of Our Time*, Beacon Press, Boston, MA.

Quarter, J 2010, *Researching the Social Economy*, University of Toronto Press, Toronto.

Renting, H, Schermer, M, and Rossi, A 2012, "Building food democracy: Exploring civic food networks and newly emerging forms of food citizenship," *International Journal of Sociology of Agriculture and Food*, vol. 19, no. 3, pp. 289–307.

Robinson, F 2010, "After liberalism in world politics? Towards an international political theory of care," *Ethics and Social Welfare*, vol. 4, no. 2, pp. 130–143.

Sia, R, Ling, J, Wu, B, Park, J, Shu, H, and Morrison, A 2013, "Women's role in sustaining villages and rural tourism in China," *Annals of Tourism Research*, vol. 43, pp. 624–650.

Su, B 2011, "Rural tourism in China," *Tourism Management*, vol. 32, no. 6, pp. 1438–1441.

Tronto, J 1993, *Moral Boundaries: A Political Argument for an Ethic of Care*, Routledge, London.

van der Ploeg, J, Ye, Z and Schneider, S 2012, "Rural development through the construction of new, nested, markets: Comparative perspectives from China, Brazil and the European Union," *The Journal of Peasant Studies*, vol. 39, no. 1, pp. 133–173.

Zhang, Q and Donaldson, J 2010, "From peasants to farmers: Peasant differentiation, labor regimes, and land-rights institutions in China's agrarian transition," *Politics and Society*, vol. 38, no. 4, pp. 458–489.

7 Farmers' markets as contested spaces

Case study of the Beijing Organic Farmers' Market

Zhenzhong Si

Introduction

In North America, the UK, Australia, and New Zealand, the last three decades or so have witnessed a dramatic growth of farmers' markets as a new direct marketing venue for food (Holloway and Kneafsey 2000; Feagan, Morris, and Krug 2004; Feagan and Morris 2009; Brown and Miller 2008; Chalmers, Joseph, and Smithers 2009). Farmers' markets are defined as "specialist markets trading in 'locally produced' products, focusing largely on food ... which is either locally grown or incorporates locally grown ingredients" (Holloway and Kneafsey 2000: 286). Geographers and sociologists have theorized farmers' markets as a part of AFNs which sit "at the intersection of the local food system ... and offers a strategic location for civic engagement with the food system" (Wittman, Beckie, and Hergesheimer 2012: 37). Most farmers' markets set up specific requirements for participation to permit local farmers growing or raising food with sustainable methods for sale to local customers. They provide a shared space for addressing environmental concerns related to food production and consumption, re-establishing the consumer–producer relations as well as facilitating renewed urban–rural linkages, and constitutes a part of the "social economy" (Hinrichs 2000; Beckie, Kennedy, and Wittman 2012; Wittman, Beckie, and Hergesheimer 2012). Farmers' markets also constitute the most visible form of local food movement and shortened food supply chains (Hinrichs 2000; Kirwan 2004; Feagan 2007). Considerable scholarly attention has been paid to interrogating the alterity (Whatmore, Stassart, and Renting 2003; Goodman 2003; Allen et al. 2003; Kirwan 2004) and the embedded localism (Winter 2003) or revalorization of "place" (Feagan 2007) of farmers' markets, alongside other AFNs, including community-supported agriculture (CSA), buying clubs, community gardens, etc. (see Jarosz 2008).

Farmers' markets as *constructed* spaces (Smithers, Lamarche, and Joseph 2008) implies that the diverse ethical values and beliefs of different actors in it, and the various forms of construction and codification of these ethical values (Chalmers, Joseph, and Smithers 2009), will all somehow shape their operation and evolution. Therefore, rather than being *coherent* and *fixed* spaces, farmers' markets are indeed dynamic, fluid, and *contested* spaces that also convey ideological collisions, discursive disputes, power struggles, and "class fragmentation and even exclusion" (Smithers, Lamarche, and Joseph 2008: 341). Existing studies of farmers' markets in the West have uncovered various tensions and conflicts. These tensions and conflicts, which reflect the contested nature of farmers' markets, render farmers' markets incoherent spaces which might impair their "alternative-ness." In this sense, these tensions are influential in determining the transformative potential of farmers' markets as a type of AFNs that challenge mainstream industrial food supply chains (see Levkoe 2011). Understanding the contested nature of farmers' markets enables us to identify potential strategies to address some of the challenges facing farmers' markets.

Although farmers' markets as contested spaces have been interrogated through different lenses, there have been very few studies examining cases in emerging economies such as China where farmers' markets are nascent initiatives. This chapter thus uses the most prominent ecological farmers' market in China—the Beijing Farmers' Market—as a case to interrogate the tensions among key stakeholders within the market space. These tensions and conflicts include (1) the dispute about the term "organic," (2) the challenges from the state in terms of the registration of the market and finding venues, the restriction of selling processed and packaged food at the market, (3) the issue of "faux-paysan," (4) the downplaying of "local," (5) challenges of communicating ethical values to customers while meeting diverse consumer needs, as well as (6) other dimensions such as the diverse farming approaches and the "old farmer" challenge.

Given that the AFN literature is largely established on Western experiences and food at these farmers' markets is more often pursued for its local provenance than its ecological feature, some tensions of the Beijing Farmers' Market we identify have not been documented previously. This chapter shows how the socio-political context of China poses various challenges for the emergence and establishment of a farmers' market. It demonstrates the significant role of the state in configuring the landscape of contestations within farmers' markets in China.

In this chapter, we will first examine the contested nature of farmers' market spaces in general in reference to experiences in Western countries. We then formulate a framework to depict the structure of contestations. Building on the structure of contestations, we apply the framework to examine the interactions, motivations and interests of various actors in the context of the Beijing Farmers' Market. In the end, we discuss the implications and significance of reading farmers' markets as a contested space.

Understanding farmers' markets as contested spaces

In Holloway and Kneafsey's (2000: 291) words, the socially constructed space of farmers' markets is an "expression of … the negotiations within networks of producers, consumers and institutions." Therefore, the tensions and conflicts among the key actors in these networks render farmers' markets contested spaces. These contestations demonstrated within farmers' markets in the West were examined in terms of various dimensions.

Tensions within the alternative space created by farmers' markets are reflected in the unbalanced power structure among actors involved. This power imbalance might unintentionally enforce "elitism and exclusion" (Chalmers, Joseph, and Smithers 2009: 324). For example, Hinrichs (2000) analyzed the social embeddedness of farmers' markets and CSAs and concluded that "tensions between embeddedness, on the one hand, and marketness and instrumentalism, on the other, suggest how power and privilege may sometimes rest more with educated, middle-class consumers than with farmers or less-advantaged consumers" (p. 295). She then concluded that poor consumers and farmers have to weigh the costs against assumed benefits from the direct social ties carefully before participating in the market (Hinrichs 2000). Swanson's (2001) interrogation of three locality-based policies illustrated that, in the operation of such locality-based food programs, the power of elites might be reproduced and thus result in polarized social classes and social exclusion. Hinrichs (2003) also pointed out that local food networks such as farmers' markets are not free of social exclusion, and locality-based projects may show evidence of the "dark side of social capital" (Schulman and Anderson 1999).

The contested nature of farmers' markets is also mirrored by the dynamic relationships among different vendors. As Beckie, Kennedy, and Wittman (2012) suggested with farmers' markets in western Canada, although competition among vendors in farmers' markets has become a driving force pushing vendors to improve and innovate, it has also resulted in a loss of vendors who could not handle the competition. As peasant identity is critical in constituting a nostalgic and authentic environment, the inherent problem of "faux-paysan" (fake farmers) (Tchoukaleyska 2013) can jeopardize the representation of farmers' markets as a nostalgic space. Tchoukaleyska's (2013) investigation of a farmers' market in France noted that farmers at the market are sometimes suspicious about other vendors not being "authentic" farmers but "resellers" of non-local produce. Smithers and Joseph's (2010) examination of farmers' markets in Canada also noted that distinguishing "authentic" farmers from "resellers" is a challenging task which may exclude some local producers from participation.

The contested nature of farmers' markets is also reflected by the interpretations of "local" in "local food systems." The locality of food sold in farmers' markets lays the foundation for the conceptualization of farmers' markets as oppositional spaces. Its feature of resistance (or, mostly, the "respatialization" of food) has been interpreted as "alternative," "conservative," "heterotopic," "defensive," and "protectionist" (Holloway and Kneafsey 2000; Winter 2003; Hinrichs, Gillespie, and Feenstra 2004; Tchoukaleyska 2013: 218). However, critical agrifood scholars argue that "local" is an outcome of struggle, and a turn to local does not necessarily imply a transition toward sustainable social and environmental relations (Eaton 2008). This is mirrored by the critiques of the "fetishized constructions of the local often present in alternative food politics" (Harris 2010: 355) and the call for a "reflexive localism" (DuPuis and Goodman 2005). "Defensive localism" (Winter 2003; DuPuis and Goodman 2005), alongside the theorization of farmers' markets as oppositional spaces, does not imply that consumer purchases at farmers' markets are necessarily intended to be oppositional or resistant (Feagan, Morris, and Krug 2004). In fact, researchers have noted the overemphasis of food "locality" in constructing farmers' markets as alternative food procurement venues (see Chalmers, Joseph, and Smithers 2009: 323). The elements and concerns of local food systems were found to "vary in their meaning, their importance and the degree to which they represent a set of absolute conditions for the participation of consumers, producers and institutional actors" (Smithers, Lamarche, and Joseph 2008: 348). The delineation of "local" and the designation of "authenticity" of food in the market are also contested and fluid in terms of what should be termed AFNs (Hinrichs 2003; DuPuis and Goodman 2005; Feagan 2007; Smithers and Joseph 2010,; Wittman, Beckie, and Hergesheimer 2012).

Tensions within the market space are also reflected by the varied motivations and values among key actors. Although farmers' markets are arguably part of an entrepreneurial and consumerist culture, in this presumed space of consumption, there are indeed negotiations over ethical values going on among organizers, vendors, and customers, such as supporting small farmers, protecting the environment, maintaining social justice, and ensuring animal welfare (see Carey et al. 2011). Nevertheless, the motivations of customers vary, rendering a resistance to certain ethical values. The importance of face-to-face communication with vendors is also questionable. Miele (2006: 351) has challenged the hegemony of "reflexive consumption" and noted that "the desire to buy organic and/or local and to have a face-to-face relationship with producers may be less important for many visitors to farmers' markets than novelty and social atmosphere" (cited in Smithers and Joseph 2010: 243). Looking into the different groups of consumers, Connell, Smithers, and Joseph (2008) highlighted the contrast between regular and non-regular patrons of farmers' markets. They conclude that the former are more concerned about organic certification, package recycling, and shopping for seasonal and local food. Chalmers, Joseph, and Smithers (2009) argued that curiosity is the factor that motivates many first-time visitors to attend new farmers' markets. Thus, although farmers' markets are conceived of as the most visible type of *local* food network, the importance of "local" and the motivations of consumers in practice vary greatly.

In summary, tensions embody power, ethical, and pragmatic conflicts among various actors. These observations in the West provide an analytical framework for the study of contestations in farmers' markets in China. It makes us wonder how farmers' markets embedded within China's specific socio-economic contexts accord with or differ from Western ones. As Ho and Edmonds (2007) noted in their examination of environmentalism in the "semi-authoritarian" context of China, there is a stronger authoritarian power but relatively weaker civil society autonomy. There is also a significant lack of trust (Wang et al. 2015; Scott et al. 2014; Sun, Meng, and Yang 2012) within Chinese society as it is transforming, according to Li (2009), from an "acquaintance society" to a "stranger society"[1] in the process of modernization. When these contextual features are translated into forces that shape the farmers' markets in China, unique tensions can be observed when examining the contested nature of farmers' markets. With these understandings in mind, in the next section, we carefully examine the various types of tensions within the Beijing Farmers' Market.

The Beijing Farmers' Market[2]

Chapter 5 provides a brief introduction of the most prominent farmers' market in China—the Beijing Farmers' Market (also known as Beijing Country Fair or Beijing Organic Farmers' Market). According to its official introduction (Beijing Farmers' Market 2013), the mission of the market is to offer a space where

> consumers and organic farmers can communicate face-to-face at the market, and get to know how healthy, safe, environmentally friendly and delicious food comes to our dining-tables. Customers can also build relations with farmers at a deeper level by visiting their farms and learning about the production and environment of farm produce in person. Amid the societal context of lack of trust, the market hopes to rebuild trust with each other with these communications so that producers can get appropriate compensation while customers can have access to healthy and quality-guaranteed food at an appropriate price.

Many farmers' markets in North America limit the vendors who want to participate in the market to those who "make, bake or grow" (Wittman, Beckie, and Hergesheimer 2012: 38). As a farmers' market heavily influenced by foreign experiences, the Beijing Farmers' Market also has a set of criteria for farms that want to participate:

• no pesticides, synthetic chemical fertilizers, GM seeds used in production;

- animals not in cages, no unnecessary (therapeutic) antibiotics or hormones;
- independent small to medium size farms;
- be open and transparent, willing to communicate with customers about the production methods (where seeds, fertilizer and feed come from, pest control methods, animal living space and stocking rates, use of greenhouses, etc.), help customers to get information and protect consumer rights;
- maintain an appropriate scale, sustainable development, and management;
- be willing to work with others (sharing farming techniques and experiences with other farmers, solving problems with customers);
- for prepared foods: use no chemical additives and be prepared in a traditional way.

Although the Chinese name of the Beijing Farmers' Market is translated as "Beijing *Organic* Farmers' Market," only a few of the 30 farm vendors at the market were certified organic producers, such as Green Yard Organic Dairy and Sunlin Ecological Farm, producing chicken and eggs. Yet, lack of certification does not jeopardize the reputation of providing safe and quality food among customers as its reputation is built upon word of mouth, direct communication with farmers, and the "gate keeping" farm visits by market managers and ordinary customers. As the Beijing Farmers' Market is gaining more attention not only in Beijing but also nationwide, a significant number of farmers and food companies (more than 300, according to the market manager in 2012) were on a waiting list to join, with the hopes of being admitted into this tightly regulated venue. However, most of them would not meet the market criteria. Of the approximately 200 vendors that market managers had visited, only about one-third qualified.

Food prices at the market are several times higher than in supermarkets and wet markets, but this does not deter the enthusiastic customers. As the demand grows, the market has opened two stores in Beijing to sell the produce of their farmer vendors. One is Community Market (its Chinese name is *jishi*, which means "gathering room") opened in 2013 and the other is Just Food (its Chinese name is *jishitang*, which means "gathering hall") opened in 2016. The stores are also offices for market managers and venues for meetings and public talks that promote ethical food consumption. The direct annual sales of the market were between CNY15 and 20 million (US$2.2–2.9 million) (Jiang 2015). The tremendous success of the Beijing Farmers' Market is a testament to the demand for quality food amid the food safety crisis. The high price and customers' demand for safe food are also mirrored by the characteristics of customers, who are mainly white-collar workers, expectant mothers and mothers of young children, and elderly people in poor health. The average income of customers is said to be CNY7,000–10,000 (about US$1,120–1,600) per month, which is above average in Beijing.

Figure 7.1 A poster introducing Beijing Organic Farmers' Market.

The Beijing Farmers' Market is a place where the relations among vendors, customers, and volunteers are based on mutual respect. In order to maintain the market's reputation of selling quality food, vendors are carefully screened and the market has been organizing farm visits frequently. A few full-time workers who run the market get paid from the earnings from the "market kitchen" where they sell their homemade food, as well as from fundraising from private foundations and from fees collected from business partners such as a wine vendor who promoted his products at the market. Volunteers, many of whom were originally customers, contribute to the market in various ways. Customers help vendors to unload and pack up, and even to sell their products. They also share information online or through WeChat—the largest Chinese social network—about strategies to get the best quality products. Introduction of produce in season, articles about ethical eating and tips and recipes about home cooking are common posts on their microblog and public WeChat account.

The Beijing Farmers' Market as a contested space

The reciprocal relationships in the market do not eliminate the embedded tensions. To understand the contestations within it, we examine the tensions between key stakeholders of the market, which include market managers, vendors, customers, and the state.

Table 7.1 illustrates the analytical framework through which, in the following section, we examine the contested spaces of the Beijing Farmers' Market. In our analysis, we compare each of these tensions and conflicts with Western experiences. Except for the "challenges from external forces," which is not observed in the case study, the Beijing Farmers' Market demonstrates all other dimensions of contestations observed in the West albeit with different connotations (Table 7.1).

Amid the challenge of food safety crises, the Beijing Farmers' Market as a nascent grassroots initiative faces numerous challenges from both within and outside the market. But tensions are expressions of interactions between players. The market is no different than its counterparts in the West in terms of being an arena for a network of a variety of actors, relations, and institutions. These key actors include market vendors, customers, market managers, and the state. Other players such as academics, NGOs, and social organizers also have roles in configuring the market as a contested space. However, in order to limit our analysis to identifying key challenges, this chapter only depicts the contested nature of the Beijing Farmers' Market in terms of the relationships between four major actors (as shown in Table 7.1). We examine these tensions in the following sections, using data collected from in-depth interviews with market managers, vendors, customers, and secondary sources.

Table 7.1 The Beijing Organic Farmers' Market as a contested space

Dimensions of contestations	Actors involved	Tensions and conflicts	Comparison with the West
Power structure	Vendors and customers	Information asymmetry	Different
	Market managers and the state	Adoption of the "organic" term	Different
	Market managers, vendors and the state	Registration and legitimacy; finding venues; perceived threats to social stability; selling processed and packaged food	Different
Vendor relationships	"New peasants" (vendors) and customers	Authentic peasant farmers being subdued; the issue of "faux-paysan"	Similar
Disputes around "local"	Market managers and vendors	Downplaying of "local"	Different
Consumer motivations	Market managers, customers, vendors	Challenges of communicating environmental and social values; maintaining environmental and social goals while meeting diverse consumer needs	Different
Challenges from external forces	N/A	N/A	N/A

Contestations about the power structure

Disputes about "organic"

The contestations about the power structure involve a key stakeholder—the state. When taking the state into consideration, the landscape of contestations within farmers' markets can be totally different. This is represented by the controversial issue relates to the term "organic," which appears in the Chinese name of the Beijing Farmers' Market: *Beijing Youji Nongfu Shiji* (Beijing Organic Famers' Market). Of all the vendors at the market, only two farms are certified organic (one is an organic dairy producer called Green Yard, the other is Sunlin Ecological Farm, which sells eggs). Green Yard gets many subsidies from the local government. The market managers were keen to have certified organic food sellers at the market to demonstrate that they are not anti-certification.

A straightforward explanation for the vendors not having organic certification is that certification in China is too costly for many small farms. This situation became even more aggravated when a more stringent national organic certification standard was enacted on July 1, 2012.[3] A close examination reveals that the understanding of "organic" (following organic principles) at the market contrasts sharply with the "organic" discourse (meeting certification requirements) in the state's certification scheme. However, the market is still called "Beijing *Organic* Farmers' Market" in Chinese, which makes it highly controversial. Critics of the use of "organic" in the Chinese name of the market claim that this causes confusion and is in fact illegal.

Shi Yan, the pioneer of CSA development in China and a well-known "new farmer" selling at the Beijing Farmers' Market, did not consider it a big problem. According to a news report (Beijing News 2013), she suggested that,

> Organic agriculture is a production system within which the soil, ecosystem and humans can be sustainable simultaneously ... whether it is organic depends on various factors. But the most important thing is the transparency of information: as long as customers know the farming practices and the information is transparent and symmetrical (between consumers and producers), whatever it is called doesn't really matter.

However, one market manager clearly acknowledged this problem in our interview,

> In the beginning we did not call ourselves an organic farmers market, because we worried it would alienate some people, and that it wasn't really true. Now we call ourselves organic and we have had some

officials tell us that technically we can't use the term. So we debate the use of this term … Beijing Country Fair [the original English name of the market] was a name that stuck.

(Interview with one market manager, December 6, 2012)

In another interview (China Broadcasting Network 2012), this manager described this dispute in a different way,

Organic agriculture and organic certified products are two different concepts. You cannot say that a farmer who follows the standards of organic farming is not organic just because he/she doesn't get certified. Most of our vendors have met organic standards. They just didn't get the certification … people without an ID are still people. But we do know that if we don't have the "ID," many of our rights are hard to protect.

Another manager also expressed his concern during our interview,

It is illegal to use the term "organic" in the name of the market but it also depends on how you understand the term. My understanding is that "organic" is our goal but not the current status.… However, we will gradually change the term in our media coverage. We also want to provide a "commitment letter" to our customers. We want the customers to know that the integrity of farming is our utmost concern. If we take out the term, we might lose customers but the final success belongs to us.… We are not against the government [in terms of the different interpretation of the term "organic"]. We don't want to cause any trouble to our vendors. The government's standards are also reasonable. In the future, if we can meet that standard, we will integrate into that system.

(Interview with one of the market managers, March 9, 2013)

These responses to the critique of using the term "organic" in the Chinese name of the market indicate that the market adopted the term to reflect the high quality of the food. Their alternative understanding of "organic" is thus positioning themselves in confrontation with the certification regime. In these circumstances, the market has introduced what they call the "networking certification" (*qinggan renzheng* in Chinese), a substitution for the expensive organic certification scheme. The "networking certification" is based on the interactions between consumers and producers in which information and knowledge is shared to the fullest extent. The market has also been conducting a participatory guarantee system (PGS) (see IFOAM n.d.), which enables various actors to participate in the monitoring and inspections of market vendors. These actors include customers, other producers, market managers, organic agricultural specialists, and the mass media. It is a way to alleviate the burden of having the limited number of market managers conduct quality assurance screening.

Although the Beijing Farmers' Market is aware of the possible clash with the state's regulation about the use of the term organic,[4] and the organizers are intending to change it in the future, these experimental grassroots initiatives of establishing "trust"—inviting customers to visit farms, telling them about the farming practices—does provide an alternative to the third-party certification scheme of quality assurance (see Wang et al. 2015). However, the struggle around the discourse of "organic" has the potential to jeopardize the market's official legitimacy.

Interacting with the state and operational challenges

The state's role in shaping the landscape of contestations within farmers' markets in China is also demonstrated by the several other challenges from the state's indirect confrontation with the Beijing Farmers' Market. This more direct encounter with the state generates a lot of problems, which mark the market as a space of contested practices. Tension in this encounter revolves around issues of registering the market as a not-for-profit entity, finding appropriate venues for the market, avoiding suspicions of social unrest, and selling processed and packaged food.

The first prominent challenge for the market managers is that it is difficult for the Beijing Farmers' Market to be officially registered as an NGO. In China, the current policies that regulate social organizations require NGOs to find a supervisory entity (usually a government department) before registering with the Civil Affairs Department of the local government (Ho and Edmonds 2007). Although there has been a recent trend to loosen this requirement, things have not changed substantially. It has proven to be a major challenge for not-for-profit entities who want more freedom in organizing activities as they have to get approval from the local authority. Local authorities will intervene in their activities to prevent troubles that they perceive might emerge. The restricted regulatory environment for not-for-profit organizations forced the market to be registered as a "social enterprise" in 2013, a compromised identity in-between a not-for-profit organization and a for-profit enterprise.

The difficulty in finding venues is the second operational challenge. This is mainly because of the large size of gathering (usually 1,000–3,000 people) that the market attracts to public spaces, which can be perceived by the government as a potential threat to social stability. In our interviews, the market managers expressed their concern about the risk of the market being called off or banned. This had happened on at least one occasion. It is also a huge challenge to deal with different government departments who can all intervene in the operation of the market.

We are cautiously balancing the influence of the market and the risk of attracting too much attention of the government: we try to avoid inter-acting directly with the administrative agencies at the lowest govern-ment level: police stations, the administrative department of industry and commerce and the city administration department. Our principle is to be very cautious when operating the market, but to keep a high profile when promoting it in the media.

(Interview with one market manager, March 9, 2013)

To cope with the potential risk posed by the local authority, the market managers tend to work with event organizers to determine venues or directly work with shopping malls or international schools that may benefit from hosting the market. (The market attracts larger-than-normal cus-tomer flow for these malls and gives publicity to them). This allows the market organizer to avoid the energy spent dealing with different govern-ment departments for necessary permits and also enables it to function as a business partner, which is less politically sensitive.

A third challenge from the state is the regulation about selling processed and packaged foods. According to the food quality governance regulations in China, producers are not allowed to produce processed and packaged foods without a production permit. In addition, their products should acquire a "Quality Safe" (QS) permission and have the QS logo on their package before entering the market (see Wang and Desmeules 2013). This creates a barrier for artisanal food sellers at the market. For example, a farmer who sells home-ground flour has to sell it loose because he is not allowed to package the flour into pre-weighed bags. A farmer at the Shang-hai *Nonghao* Farmers' Market had his packaged flour and rice confiscated because of this.

Contestations about vendor relationships

The contestation associated with "faux-paysan" (Tchoukaleyska 2013) is also observed in the Beijing Farmers' Market. Photographs showing farming scenes, naming products as "earthy" (*tu*) and "traditional," describing prod-ucts as "tasting like the food in your childhood," compose images of the "old times" for customers at the market. This vivid portrayal of the "old times" denotes the market as a nostalgic space that draws customers from across the city to the market and thousands of followers to their Weibo account. However, these "socio-politically conservative notions of place and identity" (Holloway and Kneafsey 2000: 294) are highly contested given that very few vendors in the market are actually Chinese peasant farmers. The driver of this problem, quite different from that in the West, is closely related to the deni-gration of farming and peasants within the Chinese society.

Traditional Chinese peasants have been perceived by the state and the general public as a social group with low *suzhi* (Chinese term for population quality)[5] and social status and thus a group that needs to be "civilized" (Murphy 2004; Schneider 2015). The negative connotations associated with "peasants" are also reflected by social activist James Yen's well-known labelling of Chinese peasants—"ignorant, poor, weak, and selfish" (see Shi 2012). These notions have been further fortified with the Chinese government's celebration of the mainstream urbanization and industrialization trend which marginalizes peasants on socio-economic and political fronts. Peasants in contemporary China are integrated into the wave of urbanization in various forms, such as working in cities as migrant workers and shifting from subsistence farming to commercial farming (Hu, Xu, and Chen 2010; Huang 2011; Zhang and Donaldson 2008, 2010). However, most of them still possess the land contractual right, gained through the establishment of the Household Responsibility System in the late 1970s, in their rural hometown. The land is treated more as a "social insurance against adversity" and a security for their eventual return than a simple "source of livelihood" (Fan 2008: 94). Maintaining a close connection to their farmland is still highly valued by the majority of migrants who migrate circularly (Fan 2008).

In contrast to these traditional peasants, most of the farmers' market vendors are well-educated urbanites, and thus "outsiders," who rent land in the countryside to pursue their "agrarian dreams," as depicted in media reports.[6] They are self-titled as "new farmers" (*xin nongmin, xin nongren or xin nongfu* in Chinese) to distinguish themselves from traditional Chinese peasants. More accurately, they are agricultural entrepreneurs with strong ethical values. A typical "farmer" at the Beijing Farmers' Market is a well-educated, middle-class person, with a college or university degree. Many of them are part-time farmers. Some of them quit their jobs to start farms to pursue their "agrarian dreams." They are thus different from traditional Chinese peasants, who generally have less education, learn farming from their parents, and work on the farm from the outset. Vendors at the Beijing Farmers' Market are thus not socially embedded within rural society, which means they can leave the countryside and the farm when they choose to do so.

But where are the peasants? Our visits to some of the farms selling at the market revealed that peasants were usually hired as farm workers, while the "farmers" with whom customers would have direct contact are farm operators. These so-called "new farmers" usually also do farm work but real peasants are contracted workers who are only paid a modest salary and do not share in the profit of the farm. Our investigations found that, although they have a certain degree of autonomy in farming practices, the farm workers do not have a voice in making managerial decisions. For the farm workers, the only difference between working on an ecological farm and working on an agribusiness company farm is the differences of farming methods. These subordinated farm workers are the faces behind the market whose work is largely unknown to the customers.

This is quite a contrast to one of the original goals of farmers' markets—fostering social justice by supporting small farmers and ecological ways of production.

Contestations about "local"

Unlike farmers' markets in the West, where the connotations of "local" generate tremendous debates, contestations with "local" have entirely different connotations in Chinese farmers' markets. In contrast to the notion that "local is of high quality" in the West (Holloway and Kneafsey 2000: 292), it is interesting to see how "local" is underplayed among Beijing Farmers' Market vendors in promoting the quality of food to their customers. Customers of the market are fed a lot of messages about how healthy and safe the food is, but this information places almost no emphasis on the food being "local." One of the reasons for this is that vendors perceive customers to be seeking safe and healthy food, but not "quality" food in a more comprehensive sense which embodies local, seasonal, and other features. This perception also pertains to the direct incentive for the founding of ecological farms—a reaction to the increasing demand for safe food amid the food safety crisis. The intentional downplaying of "local" might also relate to the appreciation of imported food in the Chinese food market. Food that is not domestically produced is always regarded as high-quality or even as a luxury product sold in high-end supermarket or specialty food stores. This makes the celebration of "local" even more challenging.

The downplaying of "local" is also reflected in farms from far-away provinces being permitted to participate in the market occasionally. In the past few years, more farmers from across the country joined the market to cope with customers' demand of diverse products. In mid-2017, the market introduced tropical fruits produced by farmers from Yunnan province, more than 2,800 km from Beijing, to their customers. However, this long-distance sourcing was underpinned by a social justice rationale—to support the livelihood of ecological fruit farmers in that area. At the Shanghai *Nonghao* Farmers' Market,[7] another influential farmers' market in China, we again observed that the "local" provenance of food is less important compared to the "safety" and "healthfulness" of food in screening the vendors. When a farmer from Fujian province wanted to join the market, the organizer told him that although their top concern was the production process of the food, they did prioritize local farms. "Local" is endorsed as a "bonus," but not a prerequisite.

The underplayed localism in Beijing Farmers' Markets and other Chinese farmers' markets signals their struggle of hanging on to their original values. Together with the value contradictions between market managers and customers, it presents a remarkable challenge to the market managers in terms of achieving multiple goals, including supplying local products, meeting customer demands, and contributing to social justice for peasant farmers. Balancing the multiple goals also shapes the spatial relations embedded in the operation of the market. Farmers' markets, in this sense, are not fixed spaces of consumption within a specific foodshed, but rather dynamic spaces with changing boundaries.

Contestations about consumer motivations

Like its counterparts in the West, customers of the Beijing Farmers' Market also demonstrate diverse motivations for shopping or volunteering at the market. However, since having access to safe and healthy food is the dominant motivation, there is somewhat less variation in motives for customers of the Beijing Farmers' Market. The tension exists instead between market managers and customers. This is directly associated with the ethical foundations of this alternative venue: values underpinning the production and consumption of food. These values mainly embody protecting the environment through food choices that support small-scale farms, social justice and ethical consumerism. Although these are generally identified as common values embedded in farmers' markets in the West, interviews with the managers of the Beijing Farmers' Market revealed significant differences in China.

The Beijing Farmers' Market as a contested space of ethical values is characterized by two kinds of tensions around value systems: on the one hand, the market manager struggles to convey ethical values to customers who have very different motivations; on the other hand, the market manager finds it hard to maintain the market as an ethical space, or a place to support ethical food production and consumption, and to oppose the forces that are encroaching on the space and shifting it toward a solely commercial entity.

It is not surprising to see the difference in ethics between the market manager and customers given the market manager's connections with and living and working experiences from the West. This has enabled them to transplant in China the farmers' market as an alternative venue along with a certain set of ethical values including social justice and sustainability concerns. When asked what is the most important characteristic that distinguishes the market from other food venues, the manager, who has an educational background in the US and experience working for an alternative think tank, the Institute for Agriculture and Trade Policy (IATP), immediately identified "social justice and ethical consumerism" rather than "food quality and healthfulness." In emphasizing the meaning of organic agriculture that the market is promoting, the manager said "organic agriculture is mainly about the environment, not safe food."[8] These Western associations with farmers' markets contrast sharply with the interests of customers who purchase food at the market out of food safety concerns.[9]

This contradiction in ethical value systems leads to a certain degree of compromise: in promoting the market to the public, market managers downplay aspects they consider important and instead highlight the food quality associated with healthfulness and safety. Thus, the environmental and social concerns of market organizers are intentionally watered down in communicating with customers. This watering down of environmental ethics and other compromises reduce the Beijing Farmers' Market and push it toward merely being a space for procuring safe and healthy food, and perhaps leisure.

> For customers, food safety is the biggest motivation—I'd say for 98% of our customers that is the key reason they buy at the market. This is our window into the customer right now. Right now they don't realize what else this is about—justice, fairness, ecology … for us, being ethical and giving attention to social justice are the most important criteria. After that we are concerned about the products being organic…. But we also know we need to keep diversifying to make the market attractive to a broad group of customers. We don't want to scare them away … we are in a position of making many tradeoffs between different criteria in order to bring the consumer choices and make the market a vibrant place.
>
> (Interview with a market manager, December 6, 2012)

Another source of tension around value systems exists between the market manager, producers who want to enter the market and business partners who have profit motivations. The market has received more than 300 applications from various farms who want to join the market, but two-thirds of those farms could not meet the basic criteria. They are either not following the organic or ecological production practices or are commercial farms that seek to expand marketing channels. Before the market opened its first permanent store in 2013, it tried to cooperate with a business partner; the attempt failed in the end mainly because the partner wanted to turn a fast profit and not take the responsibility for conducting farm inspections.

In recent years, the market faces an increasingly imperative dilemma that challenges one of their major initial goals, owing to the growing sales of its online store and, opened in 2013 and 2016, two physical stores. That is, although they want to promote their vendors through alternative channels, their initial goal of running the market is to facilitate face-to-face communications and consumer education. Yet, as more customers shop at their stores, the market will be sidelined and reconfigured to a product demonstration site and the "social enterprise" itself will lose its fundamental value.

Other contestations: farming practices and the "old farmer" paradox

Even if there were no power struggles, ethical value clashes, or regulatory disputes, the Beijing Farmers' Market would not be an entirely harmonious space. Farmers at the market, who all practice ecological agriculture following organic agriculture rules and principles, still have diverse perspectives on farming techniques and approaches. From our interviews, the two most prominent debates among farmers at the market relate to the use of greenhouses and pest and weed control methods. Some farms strictly follow the rule of growing only seasonal vegetables, while other farms use greenhouses to extend their growing system. Farmers also invest very different amounts of labour for pest and weed control. Some even believe pests and weeds should be allowed to flourish, following a major ecological farming trend often referred as "natural farming" or "permaculture." Even among farmers who are more proactive in pest control, there is disagreement about the right approach to maintain soil fertility.

Another contested issue is the so-called "old farmer" challenge. Despite the market's strong inclination to support "old farmers" who have been farming for all their lives, the market managers found it very hard to educate them about ecological farming ideas. One market manager said, "common wisdom suggests that they have been farming for decades so that they should know best about sustainable ways of farming, but in practice, they don't know." This is also the reason why customers at the market, who are looking for a "modern" style of farmer and do not trust peasant farmers but trust the well-educated "new farmers" instead.

Discussions and conclusions

Agrifood studies have conceptualized farmers' markets as dynamic, fluid, and contested spaces within which interactions among various actors generate tensions and conflicts. Understanding the tensions and conflicts within farmers' markets enables us to identify potential strategies to address some of the challenges. Drawing on the notion that farmers' markets are social constructions of key actors, we sought to examine the conflicts and tensions among four major players of farmers' markets: the market managers, market vendors, customers and the state. In this chapter, we used the major dimensions of contestations observed in the West as an analytical framework to interrogate the Beijing Farmers' Market in China (see Table 7.1). These dimensions included power structure, vendor relationships, disputes around "local," consumer motivations, and challenges from structural forces. This analytical framework enabled us to anatomize the Beijing Farmers' Market in various ways to capture the nuances behind its alternative characteristic.

By examining their interrelations, we sketched out the major tensions and challenges that the Beijing Farmers' Market faces, and depicted its contested nature from multiple dimensions. These dimensions of contestations include (1) the dispute about the term "organic," (2) the challenges from the state in terms of the registration of the market and finding venues, and the restriction of selling processed and packaged food at the market, (3) the issue of "faux-paysan," (4) the downplaying of "local," (5) challenges of communicating ethical values to customers while meeting diverse consumer needs, and (6) other dimensions such as the diverse farming approaches and the "old farmer" challenge.

Although some contestations within the Beijing Farmers' Market resemble those observed in the West, there are still contestations that have not been previously documented, or widely discussed in existing studies of farmers' markets that mainly examined cases in the West. Being nascent initiatives, farmers' markets in China are subject to monitoring and regulations of the state. Therefore, the unbalanced power structure might not only exist between "educated and middle-class" consumers and producers (Hinrichs 2000). Power struggles also emerge between the state and market managers. The social-political construction of specialty and quality food sold in the market can thus face significant interventions from state authorities. Moreover, the possibility of customers being unsatisfied with vendors also deserves academic attention. There are also contestations that have very different connotations from those in the West. For example, the "local" is significantly downplayed in the market. In contrast to vendors criticizing others for being resellers rather than authentic farmers, as has been observed in the West, most vendors at the Beijing Farmers' Market are actually white-collar entrepreneurs who do not have a genuine social connection with the countryside.

Even though we framed farmers' markets as spaces of contestations, the illustration and evidence of it is still highly dependent upon specific socio-political conditions. This is partly due to the fact that power structures and institutional settings vary significantly under different circumstances, which consequently affects the configurations of farmers' markets. There is no surprise that farmers' markets in other Chinese cities face very different challenges. However, as we have argued in previous chapters of the book, the tensions within the Beijing Farmers' Market still reflect the specific economic, social, and political context of China in general.

This specificity of facts and perspectives that define farmers' markets as contested spaces prevents the possibility of generating a universal approach to reconcile the fractured space of markets. Nevertheless, exploring potential mechanism that can mitigate or eliminate some of the contestations within the space of farmers' markets is still a fascinating topic for further research. The complicated tensions and conflicts unveiled in this chapter do not represent the entire picture. We acknowledge that the Beijing Farmers' Market as a contested space has more facets than what have been documented. For example, we did not include differences in values among consumers; nor did we take into account other actors such as academics and NGOs. This also demands further studies. In the process of collecting information for this study, we observed that the internet (cyberspace) has been functioning as a communication channel through which various actors participating in the market are engaging in a more equal, meaningful, and reciprocal relationship. Given the constraints of communicating through physical space in real life, cyberspace may become a venue where the various tensions could be addressed. The online and real space of the Beijing Farmers' Market and other farmers' markets across China are still developing rapidly and are worthy of continuous scholar attention.

Notes

1 The "acquaintance society" versus "stranger society" paradigm offers an effective tool to amalgamate various transformations of Chinese society. "Acquaintance society" (*shuren shehui*) was originally proposed by Chinese sociologist Fei Xiaotong in the 1940s (see Fei 1992) to characterize the traditional rural society of China, where social networks and trust are formed by blood ties or geographical relations (people from a same region). It implies a close circle type of trust based on extents of acquaintance or *guanxi*. In contrast, "stranger society" refers to a modern society where social networks mainly happen between strangers and trust is formed by formal contracts rather than acquaintance (Zhang 2005, Guo 2010).
2 Most information in this introduction, unless otherwise specified, is drawn from two interviews with the main market manager on April 3 and December 6, 2012, in Beijing.
3 This new organic certification standard requires a higher frequency of tests of environmental conditions and chemical residues in crops, which has resulted in a sharp increase in the cost of certification.
4 Other famers' markets in China do not call themselves "organic" and thus do not have this problem.
5 According to Murphy (2004, 2), the discourse *suzhi* is an all-embracing term that refers to "the innate and nurtured physical, intellectual and ideological characteristics of a person."
6 Agrarian dreams is translated as "*nong chang meng*" or "*tian yuan meng*" in the Chinese media reports, such as Wang (2013) and the TV series of CCTV-7 called "Agrarian Dreams of the Urbanites" broadcast in November 2013. It is associated with the so-called "New Peasant Movement" where urbanites rent land and start farms in the countryside.
7 I visited the Shanghai *Nonghao* Farmers' Market on May 27, 2012.
8 Interview with one market manager in Beijing on April 3, 2012.
9 Anther major motivation is the taste of food. Customers generally conclude that food sold at the Beijing Farmers' Market has a much better taste compared to that in supermarkets (see Lian 2012).

References

Allen, P, FitzSimmons, M, Goodman, M, and Warner, K 2003, "Shifting plates in the agrifood landscape: The tectonics of alternative agrifood initiatives in California," *Journal of Rural Studies*, vol. 19, no. 1, pp. 61–75.

Beckie, MA, Kennedy, EH, and Wittman, H 2012, "Scaling up alternative food networks: Farmers' markets and the role of clustering in western Canada," *Agriculture and Human Values*, vol. 29, no. 3, pp. 333–345.

Beijing Farmers' Market 2013, "Introduction of Beijing farmers' market." Accessed at http://blog.sina.com.cn/s/blog_725ab7d40100xqdt.html on September 10, 2017 (in Chinese).

Beijing News 2013, "Disputes about the organic farmers' market." Accessed at http://finance.people.com.cn/money/n/2013/0105/c218900-20095097.html on May 14, 2013 (in Chinese).

Brown, C and Miller, S 2008, "The impacts of local markets: A review of research on farmers markets and community supported agriculture (CSA)," *American Journal of Agricultural Economics*, vol. 90, no. 5, pp. 1296–1296.

Carey, L, Bell, P, Duff, A, Sheridan, M, and Shields, M 2011, "Farmers' market consumers: A Scottish perspective," *International Journal of Consumer Studies*, vol. 35, no. 3, pp. 300–306.

Chalmers, L, Joseph, AE, and Smithers, J 2009, "Seeing farmers' markets: Theoretical and media perspectives on new sites of exchange in New Zealand," *Geographical Research*, vol. 47, no. 3, pp. 320–330.

Chang, T 2012, *From Small Groups to National Network: The Status and Prospects of Farmers' Markets in Mainland China.* Working paper for the Fourth National CSA Symposium China, November 30 to December 2, 2012, Beijing, China (in Chinese).

China Broadcasting Network 2012, "Consumers endorse Beijing organic farmers' market though not certified." Accessed at http://finance.cnr.cn/gs/201208/t20120805_510478935_3.shtml on May 15, 2013 (in Chinese).

Connell, D, Smithers, J, and Joseph, AE 2008, "Farmers' markets and the good food value chain: A preliminary study," *Local Environment*, vol. 13, no. 3, pp. 169–185.

DuPuis, M and Goodman, D 2005, "Should we go 'home' to eat? Toward a reflexive politics of localism," *Journal of Rural Studies*, vol. 21, no. 3, 359–371.

Eaton, E 2008, "From feeding the locals to selling the locale: Adapting local sustainable food projects in Niagara to neocommunitarianism and neoliberalism," *Geoforum*, vol. 39, no. 2, pp. 994–1006.

Fan, C 2008, *China on the Move: Migration, the State, and the Household.* Routledge, New York.

Feagan, R 2007, "The place of food: Mapping out the 'local' in local food systems," *Progress in Human Geography*, vol. 31, no. 1, pp. 23–42.

Feagan, R and Morris, D 2009, "Consumer quest for embeddedness: A case study of the Brantford farmers' market," *International Journal of Consumer Studies*, vol. 33, no. 3, pp. 235–243.

Feagan, R, Morris, D, and Krug, K 2004, "Niagara region farmers' markets: Local food systems and sustainability considerations," *Local Environment*, vol. 9, no. 3, pp. 235–254.

Fei, X 1992, *From The Soil: The Foundations Of Chinese Society*, University of California Press, Berkeley, CA.

Goodman, D 2003, "Editorial: The quality 'turn' and alternative food practices: Reflection and agenda," *Journal of Rural Studies*, vol. 19, no. 1, pp. 1–7.

Guo, Q 2010, "Trust systems in 'acquaintance society' and 'stranger society': A reflection of trust in China based on the trust in the West," *Sixiang Zhengzhi Gongzuo Yanjiu*, vol. 5, pp. 27–29 (in Chinese).

Harris, EM 2010, "Eat local? Constructions of place in alternative food politics," *Geography Compass*, vol. 4, no. 4, pp. 355–369.

Hinrichs, C 2000, "Embeddedness and local food systems: Notes on two types of direct agricultural market," *Journal of Rural Studies*, vol. 16, no. 3, pp. 295–303.

Hinrichs, C 2003, "The practice and politics of food system localization," *Journal of Rural Studies*, vol. 19, no. 1, pp. 33–45.

Hinrichs, C, Gillespie, G, and Feenstra, G 2004, "Social learning and innovation at retail farmers' markets," *Rural Sociology*, vol. 69, no. 1, pp. 31–58.

Ho, P and Edmonds, RL (eds.) 2007, *China's Embedded Activism: Opportunities and Constraints of a Social Movement*, Routledge, London.

Holloway, L and Kneafsey, M 2000, "Reading the space of the farmers' market: A preliminary investigation from the UK," *Sociologia Ruralis*, vol. 40, no. 3, pp. 285–299.

Hu, F, Xu, Z, and Chen, Y 2010, "Circular migration, or permanent stay? Evidence from China's rural-urban migration," *China Economic Review*, vol. 22, no. 1, pp. 64–74.

Huang, P 2011, "China's new-age small farms and their vertical integration: Agribusiness or co-ops?" *Modern China*, vol. 37, no. 2, pp. 107–134.

IFOAM (International Federation of Organic Agriculture Movements) n.d., "Participatory Guarantee System (PGS)." Accessed at www.ifoam.bio/en/organic-policy-guarantee/participatory-guarantee-systems-pgs on September 13, 2017.

Jarosz, L 2008, "The city in the country: Growing alternative food networks in metropolitan areas," *Journal of Rural Studies*, vol. 24, no. 3, pp. 231–244.

Jiang, Y 2015, "A common path for PGS of farmers' markets," Partnerships for Community Development. Accessed at www.pcd.org.hk/zh-hans/work/一个共同的农夫市集的pgs之路 on July 17, 2018 (in Chinese).

Kirwan, J 2004, "Alternative strategies in the UK agro-food system: Interrogating the alterity of farmers' markets," *Sociologia Ruralis*, vol. 44, no. 4, pp. 395–415.

Levkoe, CZ 2011, "Towards a transformative food politics," *Local Environment*, vol. 16, no. 7, pp. 687–705.

Li, J 2009, "China's evolving 'double-track' socio-legal system in conflict resolution," *Peace and Conflict Studies*, vol. 16, no. 2, pp. 1–16.

Lian, Q 2012, "'Taste in my childhood': Beijingers love to shop at 'organic' farmers' market," *China Economic Times*. Accessed at http://jjsb.cet.com.cn/articleContent2.aspx?articleID=138596 on October 7, 2013 (in Chinese).

Miele, M 2006, "Consumption culture: The case of food." In: Cloke, P, Marsden, T, and Mooney, P (eds.), *Handbook of Rural Studies*. Sage, London, pp. 344–354.

Murphy, R 2004, "Turning peasants into modern Chinese citizens: 'Population quality' discourse, demographic transition and primary education," *The China Quarterly*, no. 177, pp. 1–20.

Schneider, M 2015, "What, then, is a Chinese peasant? Nongmin discourses and agroindustrialization in contemporary China," *Agriculture and Human Values*, vol. 32, no. 2, pp. 331–346.

Schulman, MD and Anderson, C 1999, "The dark side of the force: A case study of restructuring and social capital," *Rural Sociology*, vol. 64, no. 3, pp. 351–372.

Scott, S, Si, Z, Schumilas, T, and Chen, A 2014, "Contradictions in state- and civil society-driven developments in China's ecological agriculture sector," *Food Policy*, vol. 45, no. 2, pp. 158–166.

Shi, Y 2012, "De-labeling: The romantic imagination of big farms and small farms," *Green Leaf*, vol. 11, pp. 37–43 (in Chinese).

Smithers, J and Joseph, AE 2010, "The trouble with authenticity: Separating ideology from practice at the farmers' market," *Agriculture and Human Values*, vol. 27, no. 2, pp. 239–247.

Smithers, J, Lamarche, J, and Joseph, A 2008, "Unpacking the terms of engagement with local food at the farmers' market: Insights from Ontario," *Journal of Rural Studies*, vol. 24, no. 3, pp. 337–337.

Sun, F, Meng, Y, and Yang, A 2012, "The construction of three latitudes in social trust mechanism," *Journal of Jiangxi Agricultural University (Social Sciences Edition)*, vol. 11, no. 1, pp. 131–135.

Swanson, LE 2001, "Rural policy and direct local participation: Democracy, inclusiveness, collective agency, and locality-based policy," *Rural Sociology*, vol. 66, no. 1, pp. 1–21.

Tchoukaleyska, R 2013, "Regulating the farmers' market: Paysan expertise, quality production and local food," *Geoforum*, vol. 45, pp. 211–218.

Wang, J and Desmeules, C 2013, "Food safety in China: From a regulatory perspective." Accessed at www.nortonrosefulbright.com/knowledge/publications/76080/food-safety-in-china-from-a-regulatory-perspective on October 7, 2013.

Wang, RY, Si, Z, Ng, CN and Scott, S 2015, "The transformation of trust in China's alternative food networks: Disruption, reconstruction, and development," *Ecology and Society*, vol. 20, no. 2, article 19.

Wang, X 2013, "Nong Chang Meng of the middle-class: The 'new peasant movement,'" *Qilu Weekly*. Accessed at www.qlweekly.com/News/CoverStory/201303/027483.html on November 13, 2013 (in Chinese).

Whatmore, S, Stassart, P, and Renting, H 2003, "Guest editorial: What's alternative about alternative food networks?" *Environment and Planning A*, vol. 35, no. 3, pp. 389–391.

Winter, M 2003, "Embeddedness, the new food economy and defensive localism," *Journal of Rural Studies*, vol. 19, no. 1, pp. 23–32.

Wittman, H, Beckie, M, and Hergesheimer, C 2012, "Linking local food systems and the social economy? Future roles for farmers' markets in Alberta and British Columbia," *Rural Sociology*, vol. 77, no. 1, pp. 36–61.

Zhang, K 2005, "Trust in the historical coordinate: On three kinds of trust in history" *Social Science Research*, no. 1, pp. 11–17 (in Chinese).

Zhang, Q and Donaldson, J 2008, "The rise of agrarian capitalism with Chinese characteristics: Agricultural modernization, agribusiness and collective land rights," *The China Journal*, no. 60, pp. 25–47.

Zhang, Q and Donaldson, J 2010, "From peasants to farmers: Peasant differentiation, labor regimes, and land-rights institutions in China's agrarian transition," *Politics and Society*, vol. 38, no. 4, pp. 458–489.

8 Promising community organizing in China's AFNs

Theresa Schumilas

Alternative Food Networks (AFNs) in North America and Western Europe are taking the shape of complex hybrids of civil society and market-based activities and organizations that are variously active in processes of social, economic, and ecological change. The degrees to which they are trans-formative is debated as scholars and practitioners alike acknowledge the need to move beyond individual approaches to change through collectiv-ized action or "food citizenship." This action and discourse in the West is based in a long history and culture of a civil society as something that is distinct from the state and the market. The situation in China is remark-ably different in that there is no historic separation between the individual and the state, and the degree to which a new independent civil society is emerging is contested. This discussion of food citizenship leaves us with a key question. If the ability of these civic networks to influence broader systems and tackle structural injustices relates to governance and to the ability to form alliances within and across civil society organizations, what possibilities for such transformative change exist in contexts, like China, where such institutions are lacking?

In this chapter, I look at the ways in which China's AFNs seem to be moving beyond instrumental market relations in organized strategies toward food system transformation. I note that, similar to their Western sisters, Chinese AFNs can be blind to privilege and perpetuate some of the very injustices they seek to transform. Yet, I argue that using inclusive and reflexive processes, participants are building diverse networks that hold transformative potential. In contrast to AFNs in North America, however, in the context of pervasive uncertainty of an authoritarian state, Chinese AFNs are developing a more subtle repertoire of community organizing tactics and building connections to broader emancipatory spaces of global social justice movements.

However, first I offer a glimpse into China's political philosophy to describe the context for this nascent community organizing.

> Our leader brilliantly displayed his sagely prowess. In place of oppres-sion he ruled with gentleness and millions of people gave him their

hearts.... And then heaven sent no disaster. The spirits of the hills and rivers were tranquil and the birds and beasts, the fishes and tortoises, all enjoyed their lives according to their nature. But the descendants of these kings did not follow their example, and great heaven sent down disaster.... When the hungry go without food the people become unruly.

(25 Mencius, Book I, Part II, ch. 4, verse 6)

People cannot earn a living farming anymore in China. There is no honour in growing food.[1]

In China today we have enough food to eat, but what we have is not safe to eat. People are worried about feeding it to their children. It is a new kind of famine.[2]

The first quote above is an excerpt from the "Mandate of Heaven," an ancient story from the Zhou dynasty (eleventh century BC), later elaborated by Mencius and taught to every Chinese child since pre-Confucian times. It is a story about (what we in North America might call) food security, or perhaps even food justice, and the moral authority of leadership. As the story goes, a leader's mandate to rule is given by Heaven (versus a blood line or by the voice of the people). The source of legitimacy to rule is vague (Heaven), but the story is clear about how to maintain the legitimacy of leadership. To maintain this mandate, the ruler needs to ensure the harvest is secure and the peasantry is satisfied. According to the story, food insecurity is a cause for rightful rebellion. Linking governance with people's right to subsistence and food security has remained the basis of Chinese political philosophy for over 2,000 years.

In the past 50 years, China has almost miraculously transitioned from experiencing the world's worst famines to becoming the world's largest food market[3] and, as the story goes, the rulers have maintained their mandate of Heaven. But now China seems perched at a crossroads. A food safety crisis has gripped the country for two decades now and the state has been unable to address the people's concerns. There is a growing inequity between rural and urban people and millions of rural peasants have abandoned all hope of earning livelihoods from agriculture and are turning to driving taxi cabs in the city or working in village factories. This has left old people to farm in the countryside on land, which is both ecologically fragile, after decades of being pumped up by synthetic fertilizers and pesticides, and politically vulnerable under a state hungry for land to fuel its economic growth and meet its food security goals. The social and ecological costs associated with China's economic "miracle" are turning out to be extensive.

The subsequent quotes above, from volunteers interviewed for this research, illustrate the frustrations with a state that seems to be neglecting its responsibility to subsistence ethics in the social contract described by the mandate of Heaven story. While the meaning of subsistence may have changed to include food quality in addition to sufficiency, the symbolism of the mandate of Heaven story remains present in examples of urban and rural resistance in present-day China (cf. Perry 2008). Indeed, breaches in the "social contract" suggested by the story underpin the emergence of the new and diverse forms of food procurement relations that we have described in this volume as AFNs.

Organizing in a context of pervasive uncertainty

Since the 1980s, the liberal democratic notion of the separation of state and individual has grown in Chinese society and a non-government sector has exploded to tens of thousands of NGOs (Hsu 2011) in fields as diverse as education, environmental health, housing, and poverty alleviation (Spires 2011). However, scholars hasten to add that the interpretation of this expansion needs to be understood beyond mere numbers and that the Chinese understanding of NGO and civil society remains distinct from that of liberal democracies in the West, so there are many questions about what exactly "counts" as an NGO. NGOs in China are typically characterized by alliances with government versus independent institutions (Hsu 2011). In its overarching goal of maintaining harmony, the Chinese state routinely places restrictions on for NGOs with desires to resist state directives. In the face of rhetoric about "small state, big society," NGOs in China work near a "hazy, shifting boundary" (Stern and O'Brien 2012: 3), where mixed signals about what is permitted are common (2012: 3). Heilmann and Perry (2011) refer to this process as "guerrilla policy making" characterized by "continual improvisation and adjustment" that creates a climate of "pervasive uncertainty" for those challenging the state (p. 12). Indeed, advocacy on sensitive issues or use of particular tactics are more likely than others to land an NGO in trouble. For example, the state opposes actions that focus on demands for rights, or resistance that seems to be building cross-class or cross-locality alliances (Bruun 2013; Stern and O'Brien 2012).

Rules-based resistance (Perry 2008) and embedded activism (Ho and Edmonds 2007) are forms of contention that enable people to take action in more prudent and subtle ways in an authoritarian context. They involve repertoires of subtle actions operating near the edge of what is authorized but without "crossing the line." Indeed, if done in a habitual way over time, these forms of resistance can wear at the legitimacy of a system in the long run. The reason for the Chinese state's restrictions on large public gatherings (even, for example, the farmers' markets we studied) is their recognition that these assemblies can evolve into overt or confrontational styles when different individuals and small groups have the opportunity to "join up" ideas, grievances, and experiences; as yet, however, there is an absence of information on what I could consider food-related resistance in

China. This analysis is an early contribution to this nascent scholarship. The analysis proceeds as follows. First, to avoid over-romanticizing resistance demonstrated in these networks, I draw on the social justice critique of Western AFNs to reveal ways in which these networks are blind to peasant justice issues. I counter this with an analysis of the ways in which these AFNs are using reflexive and inclusive approaches, coupled with a diverse range of community organizing and subtle resistance strategies. I conclude with an illustration of how China's AFNs hold promise for transformative change through their connections to both endogenous and global justice movements.

Social justice critique

AFNs can be blind to justice issues and perhaps unintentionally exclude people in marginalized positions whom they seek to empower. Indeed, much of the critique of structural inequalities that result from "othering" is mirrored in my observations of Chinese AFNs. While the growing interest in the social economy and interpersonal relations of care (see Chapter 6) illustrates ways in which China's AFNs are trying to construct more fair relations with peasants, it would be simplistic to suggest that these nascent networks have managed to challenge deep historical problems in their brief history. Indeed, my interviews reveal a deeply held historical distrust of peasants, which works against reconnecting with the people who grow the food in these networks.

As highlighted in the discussion of care relations in Chapter 6, China's AFNs privilege connecting to land and urban entrepreneurs who operate farms, versus to the peasants who grow the food and labour on these farms. However, it is not only the consumers in these networks who display a distrust of peasant farmers; AFN organizers and CSA entrepreneurs at times also seem to contribute to a marginalization of peasants. For some of the CSA operators in these networks, peasant farmers are simply labour, and there is no attempt to integrate them into the decision-making on the farms. When asked about the involvement of peasants in the farms, these organizers replied that the peasants had lost traditional farming skills and that they would have very little to share in planning the work on the farm. This is an interesting perspective considering peasants come from families with hundreds of years of experience working on the land while the urban people starting these CSAs are new to farming. Indeed, those CSA operators who come from urban rather than peasant backgrounds seemed blind to this othering and sometimes appeared more concerned about the availability of "cheap labour" rather than celebrating or supporting recent state policies aimed at addressing rural marginalization. As one CSA operator highlighted,

> It is hard to find workers now as the government is building factories in villages to slow urbanization ... there is little incentive for workers

to come to the city to work anymore so it is getting harder and harder
to operate a CSA.

(Interview with a CSA farmer in Beijing, April 1, 2012)

She went on to recount how she uses the services of a recruitment agency
offered by the municipal government to help her locate suitable workers.

This blindness to peasant othering extends beyond CSA operators. I
attended the Fourth National CSA Conference in Beijing to distribute
surveys and was fortunate to sit beside a young Chinese university student
who spoke English well and agreed to help me locate peasant farmers who
might complete my surveys. Despite not knowing anyone in the room, she
proceeded to point out peasant farmers to me, explaining that she could
identify them by their appearance and mannerisms even though they
appeared exactly like everyone else in the room to me. She explained that
"They are of low quality in how they walk, dress and speak—I can tell by
the way they are sitting that they are peasants from the countryside," thus
reading the *suzhi* (Anagnost 2004) of people from their bodily form,
clothes, and speech. This evaluation of peasants as being of low quality is
widespread. Even the central protagonists in the AFNs we studied, whom
by all other accounts I consider to be taking strongly egalitarian positions,
at times seemed equally blind to peasant marginalization and injustice. For
example, one of the buying club organizers explained that she procures
only from CSA farms operated by urban people and not peasant farmers
because "they are hard to inspect and monitor because they don't have the
environmental ideology."[4]

I could share many more examples of the "othering" in these networks
and the marginalization of peasants in China generally, but for this
purpose it is sufficient to say that these AFNs are largely mirroring the situation in the West and that China's AFNs can reveal social injustice based
on entrenched inherited inequities. Certainly, there are efforts to address
injustices in these networks through charitable acts. Farmers' markets use
money raised from food sales to purchase food for peasants living in poor
districts as well as to subsidize peasant farmers to attend training events
and workshops. However, these localized approaches or "band aids" do
not fundamentally challenge structural conditions or cultural discourse,
such as *suzhi*, that perpetuate marginalization. In the sections that follow,
I explore the ways in which a more transformative politics are beginning
to take shape in these networks, built on reflexive ethics and community
organizing strategies.

Evidence of reflexive justice

In the West, reflexivity, or a politics of respect, is seen as an important
style of AFNs seeking to embrace and address blindness to privilege
(Goodman, DuPuis, and Goodman 2012). By working with a strong

awareness of injustices and inequalities, networks can create an open process that guards against the risk of the privileged taking hold of and co-opting the process. Reflexive processes emphasize "becoming" versus assuming desired ends and are conscious of deficiencies and pathology possible in our actions. Reflexivity involves facing and deliberating about underlying assumptions, practices, structures, and the various possible ways of framing problems and actions. AFNs demonstrating reflexivity build collaborations as "open ended stories" (Goodman, DuPuis, and Goodman 2012: 24) rather than beginning with "like-minded" people who hold a shared view of the world.

Chinese AFNs are demonstrating a commitment to inclusive and participatory process and are trying to broadly engage producers, consumers, peasants, entrepreneurs, officials, media, and many other people into assemblages that are non-hierarchical, open-ended, and networked. One of the farmers' market organizers continued to refer to these AFNs as offering a "platform" through which people assemble, discuss, and develop initiatives, noting that,

> Production and sales connection is only a small part of our market. Every year thousands of consumers come. We know this is not enough to change the big environment. But we offer this platform to let people know more about organic and about peasant farming. Some of these people will invent new activities to put on this platform, so it will never be just a farmers' market.
> (Interview with a farmers' market organizer in Beijing, April 2, 2012)

A reflexive approach embraces the struggles inherent in bringing diverse perspectives together. This is emulated by the process one of the buying clubs used to arrive at their particular definition of organic and the choice of vegetable suppliers. The club organizer explained how they needed to bring different perspectives together and talk them through. Some of her members wanted to be guaranteed that the farms they sourced from were only using inputs produced on the farm. But the farmers in their meetings told them this would be impossible and that they needed to use manure from other farms. Meanwhile, the extension experts from the university who came to the meeting advised them that they needed to rely on chemical fertilizer or there would be no food produced for their club to procure. In the end, they arrived at a set of practices that described criteria for off-farm inputs and prohibited chemical fertilizer, and accepted that this might mean lower yields. Their story demonstrates the tradeoffs made by bringing different perspectives together in an open and reflexive process.

In several ways AFNs in China demonstrate reflexive justice in the ways in which they focus on process over vision and reflect consciously about their deficiencies. For example, when I asked one of the famers' market coordinators about the ways in which peasant farmers in particular use the "platform" of the market, she confirmed my observation that most of the people volunteering to organize the market, most of the sellers, and most of the buyers are middle-class urban residents. However, she went on to explain:

> You need to understand the situation in China about the peasant. No one trusts peasants. Most of the people who come to buy at the market would never buy their goods. We want to change this. But we have only been doing this for three years and peasants have been oppressed in China for much longer than that. We know we need to expand in numbers and build trust. After that, we don't know. We will have to talk and consider. We have already gone to farms to meet with peasants and invited them to sell at the market. If people can begin to buy directly from a few peasants in these markets they will understand that they are not dirty and backward. They are efficient and hard working. We want to change things in China, but we can only walk one step at a time and cross the river by feeling the stones.
> (Interview with a farmers' market organizer in Beijing, April 2, 2012)

This openness to ideas and commitment to participatory process is further illustrated by the way the AFN volunteers position themselves as receptive to new ideas and actions. One of the market volunteers described how sometimes people come forward with "different" ideas that at first seem perhaps a little "strange" and quite removed from the operation of the market. But, after discussion, they find a way to move forward on these ideas. She explained that often these different ideas end up revealing the "fun" side of food, noting that their orientation has been to offer celebrations with food and festivals that connect people with local art and artists. Indeed, a review of their online calendar of events, coupled with the way they are continuously featured in media accounts, suggests a vibrancy about food. She described how her original "serious" approach has changed and how she has come to embrace the celebratory aspects of their work, asking:

> Who wants to join something that is old and boring? Plus, who wants to volunteer their time in activities they don't enjoy? Of course we do this because we are having fun, and we want others to have fun too.
> (Interview with a farmers' market organizer in Beijing, April 2, 2012)

Reflexivity is a struggle and not all the encounters and debates in these heterogeneous processes conclude positively. On one of my visits, there had just been a significant disagreement between a central CSA organizer and other operators at her CSA. She felt they were moving more toward a business approach and focusing on production and member engagement and that they were losing sight of the underlying marginalization of peasants that drew them to start the CSA in the first place. The struggle was not resolved amicably, and the tension was obvious in several of the meetings I attended. In the end, she moved on to remain involved in the network through a new CSA that experiments with new ways of empowering peasant farmers. Of course, the process has a positive character as well. The concern was "tabled" in a way in which everyone was allowed to "save face," while the elephant in the room was at least named.

These examples reveal producers and consumers in these networks as self-aware, ethical actors who are actively constructing these networks as communities of practice. It is worth noting that all of the interviewees I spoke with, regardless of their role in the network (farmers, consumers, organizers), used the collective pronoun "we" in describing involvements, suggesting a feeling of "being in common" with others. These and other examples depict the struggle in these AFNs to build a politics that expands opportunities for peasants and others through attention to reflexive practice. They demonstrate inclusive if not difficult dialogue that is attempting to bring together multiple perspectives, and the challenges in doing so. English literature on the importance of reflexivity to social justice describe this process as "unfixed" or "dry eyed about ideals" (Goodman et al. 2012: 156–157). Such reflexivity is about struggling with different perspectives and options that arise from bringing diverse groups to the same table, rather than bringing like-minded people together. In essence, these responses demonstrate how Chinese AFNs are trying to be simultaneously instrumental and egalitarian. I see an ideology of the market and blindness to class inequality as well as reflexivity and a commitment to inclusive open process.

Nuanced community organizing tactics

The preceding discussion describes a way of working evident in these networks. In the following sections I look at the particular resistance practices or repertoire of Chinese AFNs, or how dissent is articulated. I observe that in general these are often subtle strategies grounded by a rules consciousness (Perry 2008). Some of these strategies parallel what I might call "community organizing" strategies of AFNs in the West and thus they are familiar to us. Yet, we need to remind ourselves of the context of pervasive uncertainty in China which these actions are situated. Indeed, operations at one of the farmers' markets in this research was shut down by the state a few months prior to my interviews because too many people would be gathering in a location close to where a state assembly was going to be held. When I asked one of the market organizers if she ever worried about state repression, she smiled and told me that many of her co-volunteers suggest that she should be worried and tell her to "hide your ambition for fighting against the system." With a demonstration of remarkable strength, she responded "Hide? Why? I have done nothing wrong. I am simply living in this world."[5]

Slogan adoption

For AFNs in the West, efforts to advocate for policy changes are a central strategy for transforming food systems (Koc et al. 2008; Lamine et al. 2012; Renting et al. 2012). In China, of course, there are limited opportunities to participate directly in such processes. Instead in these AFNs we see that advocacy takes more subtle forms.

Throughout interviews and in blog posts, there was a continual reference to, and adoption of, government slogans and rhetoric, seemingly at every available opportunity. Indeed, it was rare that an interview concluded without me making note of a government slogan. Throughout the period of this research, two slogans in particular were embraced and extensively shared within AFN communications. The phrase "ecological civilization" was announced in a speech of the Sixteenth Party Congress in 2005 by Wen Jiabao, and re-confirmed in 2007 at the Seventeenth Party Congress by Hu Jintao. The phrase "beautiful China" was introduced as a central state slogan by Xi Jinping in April 2013. I spent quite a lot of time trying to get people to talk about the meanings behind these often-used phrases but this proved difficult. Interpreters simply used the phrase to explain the phrase and indicated that this was the state's direction. Finally, one interpreter explained to me that these are slogans that really can mean whatever the state needs them to mean at any given time, noting that they "mean everything and nothing, like the famous phrase, 'with Chinese characteristics.' "[6]

I suggest that this slogan adoption is a way that activists can demonstrate support for and alignment with government food-related policy, yet criticize it at the same time. The tactic is part of the embedded activism (Ho and Edmonds 2007) approach, where alignment with political rhetoric is key to maintaining productive relationships with the state. Some AFN organizers reflected quite openly on the strategy, noting that:

> The reform policy of the country leads to the detachment of peasant from villages and we are trying to help them solve this, but some might worry about gathering of people together at the farmers' market because it could lead to unrest. It can't get too big. On the other hand, we think the government could be brought to support this. So to fit in we stay with the government and use their words so they will see us as allies.
>
> (Interview with a CSA farmer in Chongqing, November 2, 2012)

Use of the internet

Criticism of state policy is not always so muted or hidden. Indeed, in the protection of private interviews (where recording was seldom permitted by interviewees) many participants voiced open criticism of state policy. AFN organizers spoke critically about the lack of funding for organic agriculture noting that the state was only interested in funding large "dragon-head" enterprises and not helping small peasant farmers. I also heard frequent criticism of the state's policies on land compensation and the corruption involved. In particular, interviewees criticized policies that deny equal benefits to migrants in the city and environmental policies such as subsidies for chemical pesticides and fertilizers. In addition, there were many overt criticisms of the organic regulations and their enforcement. However, the most significant criticism was voiced in reference to the state's inability to ensure safe food and the uselessness of food safety regulations and corrupt enforcement. What is striking is that frustration with the state's food safety governance was raised by *every* person I interviewed, even though none of my questions directly asked about this. Indeed, the extent of this dissent took me by surprise.

This dissent was also evident in internet postings. Beyond simply a recruitment and information dissemination tool, use of the internet, in particular the Weibo microblogs I monitored, can be seen as a foundational tool of resistance where people step out from behind the sarcasm and subtleties described above. In line with Yang's (2013) recent analysis of internet contention, I found bloggers to engage with food safety issues in particularly openly critical ways. The following are a few examples of posts:

> We are tired of all the talk of food safety—it's ridiculous—every day there is a new problem and the government is doing nothing. They are irresponsible. But they have their own special food supply so they don't care about us.[7]

> There is corruption everywhere. Officials know about these problems and they accept bribes and leave the practices continue. It is embarrassing for me to say this to you.[8] I don't understand how Chinese people can do this to other Chinese people—deliberate adulteration of food—but worse than that, I cannot understand why the government does nothing. Someone should resign.[9]

These echo findings in Yang's (2013) research. He cites remarkably similar postings such as:

> There is too much talk about food safety, too much already. It's hopeless. Manufacturers still do as they like. The supervisory agencies are still absent. Common folks—just pray for your own luck.

> We don't have the safe "specially provided foodstuffs" available to the privileged. We can only toughen up our own stomachs. Perhaps eventually we will evolve into some alien forms.

Yang has conducted extensive research on Chinese resistance and the internet. He underscores the significance of these food safety responses, which may appear rather benign to those of us participating in AFNs in North America. However, Yang notes that in China's political context some of these posts may trigger large-scale social disturbances that indeed threaten regime security; as one of the bloggers he cites suggested: "If the food safety problem is still not solved in China, it will surely become the biggest problem affecting harmony and stability."

Facilitating voice

In the West, policy advocacy as undertaken by AFNs frequently involves community-based processes which "give voice" through the democratic process to diverse community members, often through grassroots research and consultation projects, which organizations then use as basis for policy advocacy (Koc et al. 2008). In China, such consultation has not been part of the ethic of developing policy. When I asked people if they had been involved in the process to create the state's organic standard, for example, several different AFN participants looked at me rather incredulously and I realized it was a naïve question. In this context, I suggest it is quite remarkable that AFNs embrace nascent community consultation processes. For example, one of the farmers' market organizers used the coincidence of my presence in China to organize a community meeting in which I could help to "encourage" AFN participants with examples of AFNs and organizing activities from Canada. Far from being simply a venue for me to present information, however, the meeting evolved into a forum where different perspectives were collated and the farmers' market volunteer prepared a document summarizing issues and themes important to China's emerging AFNs. On a subsequent visit, she showed the document to me and explained that it is their "version of your people's policy process"[10] that "starts to organize our views of what is needed in China and the work that AFNs can do."[11] In a second example, one of the CSAs conducted a fledgling study of CSAs in China, documenting how many there are, and their types, and also exploring people's motivations for joining and concerns with the dominant food system. They shared their findings online, through their CSA newsletter, and at the CSA conference in Beijing. Subsequently, this fledgling work was taken up by academics and enhanced and has become the early stages of a Chinese academic scholarship on AFNs (see Chen 2013a, 2013b).

Building guanxi

AFN volunteers invest significant time enlarging their networks by forging ties with members of other nascent civil society groups, environmental NGOs, and the media. However, the relationships with academic allies and representatives of the state seem particularly well developed. Several CSAs are connected to local government officials. For example, one operator described how a local government representative seemed quite interested in the CSA noting that,

> Even though he can offer us nothing we need ... right now he can only offer us a reduced price on fertilizer, but we don't need that.... He will still be useful to us one day, so we keep inviting him to events and we bring him food because we are cultivating *guanxi* with him.
>
> (Interview with a CSA farmer in Beijing, December 6, 2012)

For another farm, the relationship or *"guanxi"* cultivated with local officials paid off when the land they use was threatened by expropriation. She described,

> See all these apartments—a few years ago this was a village. It is now gone, and here we are left, one little farm in the middle of this.... We would be gone too, except for *guanxi* we had built with some local officials.
>
> (Interview with a CSA farmer in Beijing, April 1, 2012)

In another example, a farmers' market volunteer explained how she was building a relationship with the state's representative to IFOAM because he was influential and would be able to "assist them somehow in the future."[12] Apparently she was right. Since completing my research, I've learned that some of the farms in these networks are now working with IFOAM on developing a participatory guarantee system (PGS) of organic verification.

Cultivating relationships and building networks with academics seem to be a particular strategy with student projects and jointly organized conferences being common to several of the CSAs. For example, one quite influential academic is a strong supporter of the CSA approach and of AFNs. Professor Wen Tienjun is a previous dean of the Institute of Advanced Studies for Sustainability and the School of Agronomics and Rural Development at Remnin University and former advisor to the State Council on Rural Development. He is credited with the formulation of the foundational "three rurals" policy mentioned above and has continued as a strong advocate of positioning rural well-being in China beyond the question of agricultural production (Wen 2007; Wen et al. 2012). Dr. Wen spoke passionately about rural reform in China at both the CSA conference and the South-South development conference I attended as part of my fieldwork, and from the reaction of the audience I suggest he is seen as a kind of "academic leader" of the AFN movement and offers it some legitimacy. He is also a leading protagonist of the endogenous rural redevelopment movement known as new rural reconstruction (Day 2008; Hale 2013) that is entangled with these AFNs, as discussed in Chapter 9.

Trans-local linkages

The linkages being built by AFN participants extend beyond China and include a widening range of connections with like-minded organizations and networks around the world. This heterogeneous alliance development is precisely the kind of process that scholars argue is most provocative to the Chinese state (Heilmann and Perry 2011; Yang 2009). While the state has been tolerant toward resistance that is limited to particular locations or isolated incidents or groups with small participation, large heterogeneous linked processes are seen as a threat to the state's hegemony. I argue that these global connections are examples of "frame bridging" described by social movement scholars. Frame bridging refers to the process of building ideologically congruent discourse and practices or "frames" that join up otherwise unconnected actors. I described earlier how diverse coalitions and networks are being built by AFNs in other parts of the world. The difference in this process for Chinese AFNs, however, is that these linkages are through personal (*guanxi*) versus organizational connections in order to avoid the risk inherent in forming official and overt multi-network movements. In this way, China's AFNs, while not officially engaged in transnational movements, are positioned as portals to a wide diversity of global movements for *individuals* interested in pursuing connections.

Some of these linkages have been advanced by a strong orientation to Western models that is endorsed by the state as part of its "opening." This approach has encouraged the drawing in of knowledge, experience, and information from outside and indigenizing these with "Chinese characteristics." For example, two leading organizers in the AFNs I studied have strong affiliations with specific international NGOs. Moving beyond these particular and *official* connections, I uncovered a number of *personal* connections to global food and environmental justice movements, where there is an entanglement of relations that is difficult to unpack. Perhaps the strongest example of this is the entanglement between the New Rural Reconstruction Movement, global food justice movements, and the AFNs. Through this close association with NRR, AFNs become a portal, or a path to linkages with trans-global food justice movements, that otherwise have no official presence in China. Indeed, in research on NRR in China, Alexander Day (2008) and Matthew Hale (2013) describe these connections and illustrate the ways in which the NRR movement resonates strongly with non-Chinese movements such as the Zapatistas in Mexico, the Landless Workers Movement (MST) in Brazil and La Via Campesina, highlighting attendance at conferences, meetings and anti-WTO protests outside of China. Representatives from these global justice movements were present at the South-South conference I attended and it was clear that many AFN organizers have established personal relationships with these groups. In this way, China's AFNs open the door to participation in global justice movements while remaining under the state's radar.

Conclusion

This chapter has argued that AFNs are not simply sites of material transactions. They are also places where community is being built and hegemony is being challenged in subtle ways. Consumers and producers are
working together in these networks, in market-based as well as civil society
relations, to articulate strategies for social change. While these are on one
hand "market-based" networks, I suggest these are also networks where
"food citizens" are being enacted, decentering private needs and centering
actions for the public good.

This analysis opens up my understanding of advocacy and political
action to reveal subtle yet powerful forms of contention that occupy the
large space between compliance with hegemony and overt defiance.
China's AFNs work at a "hazy, shifting boundary" (Stern and O'Brien
2012: 3) between permitted and prohibited, and a climate of "pervasive
uncertainty" (Heilman and Perry 2011: 22). China's AFNs can be understood as remarkably reflexive networks characterized by struggle toward
inclusiveness. While on the one hand activists in these networks can be
blind to their privilege, they are also trying to address a deep historical distrust of peasants that works against reconnecting with people who grow
food. Further, given its diffuse nature, the internet provides a platform for
activists to extend their reach, offering new possibilities for community
organizing as well as voicing dissent. These AFNs draw support at diverse
scales including both endogenous rural development movements and international NGOs to build connections to global food justice movements that
have no official presence in China. Hence, while these are on the one hand
market-based networks, they can also be considered laboratories where
food consumers are becoming "food citizens" and are centring actions for
the public good and decentring their private needs.

The symbolism of the mandate of Heaven story presented at the beginning of this chapter has relevance for China's AFNs. For these activists, the
meaning of subsistence has changed and now includes food safety and food
quality in addition to sufficiency. They see the state as unable to secure
safe food, prompting the emergence of these new forms of food relations
and motivating nascent civil society organizing around food. Do AFN
activists see the food quality crisis in China as reducing the Communist
Party's authority to govern, as suggested by the mandate of Heaven story?
Certainly, the people we interviewed and the posts we monitored suggest
that people understand poor food quality to be an issue that the state is
responsible for fixing. Further, some think there is a possibility that, should
it continue, poor food quality in China and food safety problems could
trigger large-scale social disturbances and threaten regime stability. So,
whether or not the "food is in the pressure cooker" remains an open
question.

Notes

1 CSA farm operator in Beijing, April 2, 2012.
2 Buying club operator in Beijing, April 2, 2012.
3 China continues to be the world's largest consumer market for food and beverage products, with a €440 billion turnover in 2014 (see EU SME Centre 2015; Garnett and Wilkes 2014).
4 Interview with the organizer of a buying club in Beijing, April 9, 2012.
5 Interview with a farmers' market organizer in Beijing, April 2, 2012.
6 Guide and translator in Changzhou, April 12, 2012.
7 CSA farmer from outside Beijing, March 2013.
8 Peasant farmer from Shanghai, February 2013.
9 CSA farmer from Guangzhou, March 2013.
10 One of the examples we shared in our presentation to the group was of Food Secure Canada's process of grassroots organizing through its Peoples Food Policy project.
11 Farmers' market volunteer in Beijing, December 6, 2012.
12 Farmers' market volunteer in Beijing, December 6, 2012.

Bibliography

Allen, P 2010, "Realizing justice in local food systems," *Cambridge Journal of Regions, Economy and Society*, vol. 3, no. 2, pp. 295–308.

Allen, P and Guthman, J 2006, "From 'old school' to 'farm-to-school': Neoliberalization from the ground up," *Agriculture and Human Values*, vol. 23, no. 4, pp. 401–415.

Allen, P and Sachs, C 2007, Women and food chains: The gendered politics of food," *International Journal of Sociology of Food and Agriculture*, vol. 15, no. 1, pp. 1–23.

Anagnost, A 2004, "The corporeal politics of quality (suzhi)," *Public Culture*, vol. 16, no. 2, pp. 189–208.

Bedore, M 2010, "Just urban food systems: A new direction for food access and urban social justice," *Geography Compass*, vol. 4, no. 9, pp. 1418–1432.

Bruun, O 2013, "Social movements, competing rationalities and trigger events: The complexity of Chinese popular mobilizations," *Anthropological Theory*, vol. 13, no. 2, pp. 249–266.

Chen, W 2013a, "Perceived value in community supported agriculture (CSA): A preliminary conceptualization, measurement, and nomological validity," *British Food Journal*, vol. 115, no. 10, pp. 1428–1453.

Chen, W 2013b, "Perceived value of a community supported agriculture (CSA) working share: The construct and its dimensions," *Appetite*, vol. 62, no. 1, pp. 37–49.

Cheng, H 2012, "Cheap capitalism: A sociological study of food crime in China," *British Journal of Criminology*, vol. 52, no. 2, pp. 254–273.

Day, A 2008, "The end of the peasant? New rural reconstruction in China," *Boundary 2*, vol. 35, no. 2, pp. 49–73.

DeLind, L and Bingen, J 2008, "Place and civic culture: Re-thinking the context for local agriculture," *Journal of Agricultural and Environmental Ethics*, vol. 21, no. 2, pp. 127–150.

Dong, L and Tian, K 2009, "The use of Western brands in asserting Chinese national identity," *Journal of Consumer Research*, vol. 36, no. 3, pp. 504–523.

DuPuis, E, Harrison, J, and Goodman, D 2011, "Just food?" In: Alkon, A and Agyeman, J (eds.), *Cultivating Food Justice: Race, Class, and Sustainability*. The MIT Press, Cambridge, MA.

EU SME Centre 2015, *Sector Report: The Food and Beverage Market in China*. EUSME Centre, Beijing.

Garnett, T and Wilkes, A 2014, *Appetite for Change: Social, Economic and Environmental Transformations in China's Food System*. Food Climate Research Network.

Gerth, K 2003, *China Made: Consumer Culture and the Creation of the Nation*. Harvard University Press, Cambridge, MA.

Goodman, D, DuPuis, E, and Goodman, M 2012, *Alternative Food Networks: Knowledge, Practice and Politics*. Routledge, Abingdon.

Guthman, J 2008, "Neoliberalism and the making of food politics in California," *Geoforum*, vol. 39, no. 3, pp. 1171–1183.

Hale, M 2013, "Tilling sand: Contradictions of the 'social economy' in a Chinese movement for alternative rural development," *Dialect Anthropology*, vol. 37, no. 1, pp. 51–82.

Heilmann, S and Perry, EJ 2011, "Embracing uncertainty: Guerrilla policy style and adaptive governance in China." In: Heilmann, S and Perry, E (eds.), *Mao's Invisible Hand: The Political Foundations of Adaptive Governance in China*. Harvard University Press, Cambridge, MA.

Ho, P and Edmonds, R 2007, *China's Embedded Activism: Opportunities and Constraints of a Social Movement*. Routledge, Abingdon.

Holdaway, J and Husain, L 2014, *Food Safety in China: A Mapping of Problems, Governance and Research*. Forum on Health, Environment and Development (FORHEAD), Social Science Research Council, New York.

Hsu, C 2011, "Even further beyond civil society: The rise of internet-oriented Chinese NGO," *Journal of Civil Society*, vol. 7, no. 1, pp. 123–127.

Johnston, J 2008, "The citizen-consumer hybrid: Ideological tensions and the case of Whole Foods Market," *Theory and Society*, vol. 37, no. 3, pp. 229–270.

Klein, J 2009, "Creating ethical food consumers? Promoting organic foods in urban southwest China," *Social Anthropology*, vol. 17, no. 1, pp. 74–89.

Klein, J 2013, "Everyday approaches to food safety in Kunming," *The China Quarterly*, vol. 214, pp. 376–393.

Koc, M, MacRae, R, Desjardins, E, and Roberts, W 2008, "Getting civil about food: The interactions between civil society and the state to advance sustainable food systems in Canada," *Journal of Hunger and Environmental Nutrition*, vol. 3, no 2–3, pp. 122–144.

Lamine, C, Darolt, M, and Brandenburg, A 2012, "The civic and social dimensions of food production and distribution in alternative food networks in France and Southern Brazil," *International Journal of Sociology of Agriculture and Food*, vol. 19, no. 3, pp. 383–401.

Levkoe, C 2011, "Towards a transformative food politics," *Local Environment*, vol. 16, no. 7, pp. 687–705.

Li, C 2010, "Characterizing China's middle class: Heterogeneous composition and multiple identities." In: Li, C (Ed.), *China's Emerging Middle Class*. Brookings Institution Press, Washington, DC.

Lyson, T 2005, "Civic agriculture and community problem solving," *Culture and Agriculture*, vol. 27, no. 2, pp. 92–98.

Nyiri, P 2009, "From Starbucks to Carrefour: Consumer boycotts, nationalism and taste in contemporary China," *PORTAL Journal of Multidisciplinary International Studies*, vol. 6, no. 2, pp. 1–25.

O'Brien, K and Li, L 2006, *Rightful Resistance in Rural China*. Cambridge University Press, New York.

Perry, E 2008, "Chinese conceptions of 'rights': From Mencius to Mao—and now," *Perspectives on Politics*, vol. 6, no. 1, pp. 37–50.

Pratt, J 2009, "Incorporation and resistance: Analytical issues in the conventionalization debate and alternative food chains," *Journal of Agrarian Change*, vol. 9, no. 2, pp. 155–174.

Renting, H, Schermer, M, and Rossi, A 2012, "Building food democracy: Exploring civic food networks and newly emerging forms of food citizenship," *International Journal of Sociology of Agriculture and Food*, vol. 19, no. 3, pp. 289–307.

Schneider, M, "What, then, is a Chinese peasant? Nongmin discourses and agro-industrialization in contemporary China," *Agriculture and Human Values*, vol. 32, no. 2, pp. 331–346.

Schumilas, T 2014, "Alternative food networks with Chinese characteristics," thesis, University of Waterloo, Canada.

Scott, S, Si, Z, Schumilas, T, and Chen, A 2014, "Contradictions in state- and civil society-driven developments in China's ecological agriculture sector," *Food Policy*, vol. 45, no. 2, pp. 158–166.

Si, Z, Schumilas, T, and Scott, S 2015, "Characterizing alternative food networks in China," *Agriculture and Human Values*, vol. 32, no. 2, pp. 299–313.

Spires, A 2011, "Contingent symbiosis and civil society in an authoritarian state: Understanding the survival of China's grassroots NGOs," *American Journal of Sociology*, vol. 117, no. 1, pp. 1–45.

Stern, R and O'Brien, K 2012, "Politics at the boundary: Mixed signals and the Chinese state," *Modern China*, vol. 175, pp. 1–25.

Wen, T 2007, "Deconstructing modernization," *Chinese Sociology and Anthropology*, vol. 39, no. 4, pp. 10–25.

Wen, T, Lau, K, Cheng, C, He, H, and Qiu, J 2012, "Ecological civilization, indigenous culture and rural reconstruction in China," *Monthly Review*, vol. 63, no. 9, pp. 29–44.

Yan, Y 2012, "Food safety and social risk in contemporary China," *The Journal of Asian Studies*, vol. 71, no. 3, pp. 705–729.

Yang, G 2005, "Environmental NGOs and institutional dynamics in China," *The China Quarterly*, vol. 181, pp. 46–66.

Yang, G 2009, *The Power of the Internet in China: Citizen Activism Online*. Columbia University Press, New York.

Yang, G 2013, "Contesting food safety in the Chinese media: Between hegemony and counter-hegemony," *The China Quarterly*, vol. 214, pp. 335–355.

9 Rural development initiatives amid the food safety crisis

Strategies, challenges, and opportunities in the "New Rural Reconstruction Movement" in China

Zhenzhong Si and Steffanie Scott

Introduction

While more than three decades of implementation of the Reform and Opening reform in China has profoundly transformed the countryside, it has also resulted in serious problems. The countryside suffers from the loss of farm labour, the stagnation of rural livelihoods, and the deterioration of rural culture (Wen 2009). These problems resulted in the poverty of peasants, the insecurity of peasants' rights, the poor condition of rural infrastructure, the low viability of the agriculture sector, the disparity between cities and the countryside, and other associated social economic problems (Yeh, O'Brien, and Ye 2013; Yan and Chen 2013). However, state-led developmental approaches (the agroindustrial model) in order to revitalize the countryside as an integral part of the market economy failed to address some of these social and cultural concerns (Pan and Du 2011a, 2011b).

It was within this socio-economic context that alternative rural development initiatives emerged in China as critical reflections on mainstream "modernization." The most prominent initiative is the New Rural Reconstruction Movement (NRRM) (*xin xiangcun jianshe yundong* in Chinese) that emerged in around 2003. It follows the values and sentiments of the Rural Reconstruction Movement (RRM) that took place in the 1920s and 1930s. "Civilian" education (in contrast with "elite education"),[1] cultural activities, and capacity building for self-organization were several key components of the RRM, which are revived in the NRRM. The major ideas of the NRRM include critiques of elite culture, knowledge and related theories (e.g. neoclassical economics), focusing on cultivating peasant status and subjectivity, connecting intellectuals with the rural masses, reconstituting rural–urban relations, experimenting with rural education reforms, and improving rural health care conditions, among others (Pan and Du 2011a, 2011b). In addition to their ambitious agendas, the NRRM creatively added another important layer, the ecologicalization of agricultural production, to address the critiques of peasants in generating food safety problems.

Although food safety and food system transformations in China are not the only issues that the NRRM concerns, food is indeed a critical factor that has shaped its developmental trajectory. The synergies between the rural reconstruction initiatives and the growing public concerns of food safety catalyzed ecological agriculture to become a prominent rural development instrument. Amid the broad context of food safety crisis that has been discussed in the previous chapters, the NRRM plays various and vital roles in fostering the development of ecological farming and AFNs, especially CSAs in China. They also interact proactively with other players such as environmental NGOs, government, public media, and social activists to magnify the impacts of their efforts. In this sense, the ecological aspect of the new rural development approach later turned out to be an unexpected entry-point for the NRRM to step into the vortex centre of the power dynamics revolving around food safety.

The NRRM's principles, strategies, and approaches all differ significantly from the state-led agroindustrial model of rural development. It falls rather into the "rural development" model, which was theorized by geographer and sociologist Terry Marsden and several others (Marsden, Banks, and Bristow 2002; Marsden 2008; Marsden and Sonnino 2008) as an agrarian-based rural development trajectory that discards the "agroindustrial" and the "post-productivist" understandings of the rural space. The "rural development" model employs short food supply chains as effective tools to counter the large-scale industrialized food value chains and centres on agriculture production to achieve rural sustainability goals (Marsden, Banks, and Bristow 2002, Marsden 2008). Despite the increasingly important roles civil society plays in China, the state still plays the leading role. Grassroots initiatives thus have to seek support from the masses on the one hand and an unconfrontational or harmonious relationship with the state on the other. In this chapter, we turn our focus from alternative food networks to the NRRM that to a great extent underpins their development.

The emergence of the New Rural Reconstruction Movement

In 1996, Wen Tiejun, an economist working for governmental agencies, conceptualized the many problems plaguing rural development in China as *sannong wenti*, which means "the problems of peasants (*nongmin*), rural society (*nongcun*), and agriculture (*nongye*)" (see Ahlers and Schubert 2009; Day 2008, 2013a, 2013b).[2] Since Wen proposed this term, *sannong* has entered the policy circle and became a term that encapsulates the complex socio-economic challenges that China's countryside has been facing. "The peasant" has been brought back to the centre of the realms of politics, economy and development in China (see Day 2008, 2013a).

The NRRM was initiated by academics and social activists in the early 2000s to tackle the *sannong* challenges. It has been led by concerned academic researchers (professional intellectuals; see Hao 2006) and NGO leaders focusing on rural China. NRRM leaders suggest that the crisis in rural China cannot be treated merely as an economic or agriculture production issue but rather as a social and cultural issue that requires the reconstruction of social life by means of cultural and cooperative reorganization (Day 2008, 2013a). Thus, the NRRM boldly aims to rebuild collective social relations in rural China for peasants to defend themselves against globalization, marketization, and capitalist-consumerist values (Day 2008, 2013a, 2013b; Jacka 2013). In 2003, Wen facilitated the foundation of "James Yen Rural Reconstruction Institute," a key symbol of the revitalization of the RRM. The initiative was later joined by other activities such as the establishment of Beijing Liang Shuming Rural Reconstruction Center and Little Donkey Farm. Besides Wen Tiejun, He Xuefeng, based at the Huazhong University of Science and Technology, was another influential scholar who carried out rural reconstruction experiments in Hubei province (Pan 2012; Thøgersen 2009; Day 2013b).

Among the variety of work that the NRRM has been doing, ecological agriculture is one important component that has attracted much public attention. The synergies between the rural reconstruction initiatives and the growing public concerns regarding food safety after 2008 placed ecological agriculture in the spotlight as a prominent rural development instrument. Within the broad context of food safety crisis, the NRRM activists played various vital roles in fostering the development of ecological farming and alternative food networks, the most prominent of these being CSA farms.

Existing literature about the NRRM, although very limited, has provided both empirical and theoretical examinations of the movement. However, little attention has been paid to the most recent food safety crisis, which has profound implications for the development of the NRRM. In this chapter, we unveil how the NRRM, positioned in the middle of the state–society dichotomy, negotiates not only with the state's authority but also society's expectations. Based on in-depth interviews with key stakeholders in the NRRM, this chapter thus interrogates the movement in the broad socio-political context of China and provides a detailed analysis of how a social movement survives within a complicated socio-political context, particularly the heightening food safety anxiety. Understanding the NRRM sheds light on our discussion of the bottom-up initiatives of agriculture ecologicalization.

Existing studies of the New Rural Reconstruction Movement

Despite the significance of the NRRM in rural development in China, there is limited literature in English documenting its origin, strategies, practices, challenges, and opportunities. Existing studies, although not many, have analyzed the historical evolution of the NRRM and its connections with the RRM in the 1930s (Day 2008, 2013a, 2013b; Thøgersen 2009,; Yan and Chen 2013), the anti-modernization sentiment within the movement (Wen and Lau 2008; Wen et al. 2012), the impacts of the NRRM on the restoration of collective rural culture (Renard and Guo 2013), and peasant organization and rural self-governance (*zizhi*) (Day 2013b). By situating the movement in Chinese socio-political and historic context, critical sociologists and agricultural economists also unveiled the neglection of gender inequalities (Jacka 2013) and the various inherent contradictions within the movement (Pan and Du 2011a).

Papers from a special issue of *Chinese Sociology and Anthropology* in 2007 detailed some early experiments of the NRRM in China (Day and Hale 2007; Wen 2007; He X. 2007; Tan 2007; He H. 2007; Qiu 2007). Authors interrogated the various problems facing rural China, pointed out the key objectives of the movement such as revitalization of rural culture and reconstruction of peasants' subjectivity,[3] and summarized pragmatic experience of rural reconstruction initiatives. Another special issue of *Chinese Sociology and Anthropology* in 2008 embodies discussions and debates around rural governance, particularly villager self-governance, from another influential new rural reconstruction group led by He Xuefeng—the Central China School of Rural Studies (*huazhong xiangtu pai* in Chinese) (He 2008; Chen 2008; Wu and Li 2008; Ying 2008; Wu et al. 2008).

By examining the NRRM through both empirical cases and theoretical reflections, these studies provided valuable information for the understanding of alternative rural development initiatives in China. Yet, the NRRM has not been examined against the backdrop of the recent food safety crisis in China. This is mainly because the NRRM had emerged and proliferated before the food safety crisis. With the heightening food safety crisis, the strategies and primitive foci of the NRRM have dramatically changed in the past few years. Under this circumstance, previous studies of the NRRM provided limited help in understanding how the NRRM strives to survive the political pressure while meeting society's demands. This chapter thus places the NRRM in the broad socio-political context of China, where it strives to make an impact. It examines the challenges and opportunities for the NRRM by linking it with the food safety crisis and the development of AFNs in China.

The origins and practices of the New Rural Reconstruction Movement

Since the late nineteenth century, China has experienced various shifts in development thought. The "radical versus conservative" paradigm, although being criticized for its dichotomy, offers an analytical framework to characterize these far-reaching shifts (see Yu 2006). Radicalism is characterized in China by the tendency to totally repudiate traditional developmental ideas and to fully embrace Western thoughts instead. In contrast, conservatism calls for a revitalization of tradition and traditional thoughts. The historical period around the "May Fourth Movement" in 1919 marked the first culmination of "radical" thoughts in the twentieth century, followed by a decline of radicalism and a rise of conservatism (Xu 2000). The rise of conservatism was fostered by the rigorous reality of the devastated rural economy and culture after World War I. According to Wen (2009), in the 1920s there was a rapid industrialization period in China that was built upon the exploitation of the rural economy. The migration of peasants to cities and the capitalization of the agriculture sector resulted in the decomposition of the subsistence rural economy and the decline of rural society.

It was within this complex "radical-conservative" struggle that the RRM emerged (see Yan and Chen 2013). This movement, led by social activists such as Liang Shuming, Yan Yangchu (also known as James Yen) and Lu Zuofu, expanded to more than 600 organizations in the 1920s and 1930s (Pan and Du 2011a). It aimed to revitalize the rural economy and rural culture through various means (see Pan 2012). The revitalization efforts were represented by rural reconstruction experiments, led by social activists, that focused on various aspects of rural life, including building autonomous institutions, job training, civilian education, traditional culture education, and public health improvement. These experiments flourished across the country in the 1920s and 1930s and constituted a remarkable social movement in the history of China's rural development (Pan and Du 2011a; Pan 2012; Yan and Chen 2013).

However, this far-reaching movement was generally perceived as a movement of "amelioration," in contrast with the more radical movement of "revolution" (Guo 2009; Liu 2008; Yan and Chen 2013). The perceptions were that the RRM, although widely embraced, did not seek a systematic transformation of the fundamental orders and institutions and thus had a limited capacity in solving the profound social and economic problems that confronted rural China at that time (Pan 2012; Day 2008; Yan and Chen 2013). The success of the RRM has also long been contested, even by its leaders such as Liang Shuming, who admitted the severe limitations and difficulties of the RRM (Liang 1936; see also Yan and Chen 2013). Its prosperity was soon halted by the outbreak of the Anti-Japanese War in the late 1930s, when the attention of the state and civil society shifted from internal treatment of the destitute countryside toward the external threat of sovereignty (Day 2013b; Yan and Chen 2013).

More than 60 years later, rural China faced severe challenges again after decades of rapid urbanization and industrialization, which hollowed the countryside by constantly extracting key human and natural resources and capital (Shi 2012; Yeh, O'Brien, and Ye 2013). This challenging condition prompted responses from both the state and civil society to launch a new round of rural development initiatives. In the early 2000s, the threatened agriculture-dependent livelihoods of peasants as well as the politically sensitive income disparity and social gap between cities and the countryside attracted the attention of the central government, owing to a well-known open letter written by a local official, Li Changping, to Premier Zhu Rongji (Day 2008). *Sannong wenti*, as expressed in Li Changping's letter, soon became a key focus of a new set of national policies marked by the introduction of agriculture subsidies in 2004 and the abolishment of agricultural tax in 2006. In the fifth plenary session of the Sixteenth Party Congress in October 2005, the central committee of the Chinese Communist Party (CCP) adopted the term "building a new socialist countryside" (*jianshe shehui zhuyi xin nongcun*) in the guiding document for the Eleventh National Five-Year Plan (2006–2010). Since then, throughout the Hu–Wen administration until 2013, "new socialist countryside construction" (NSCC) serves as an all-embracing term that embodies various state efforts in developing the countryside, with agriculture modernization as the top priority. These efforts cover the modernization of the agriculture sector; protection of grain price; improvement of education, medical care, and transportation infrastructure; and the beautification of the countryside, as well as enhancement of villagers' self-management capability (Central Committee of the CCP 2005).

Paralleling but prior to the state's NSCC agenda, the legacy of RRM was "salvaged"[4] by a group of intellectuals led by renowned economist, Wen Tiejun, who proposed and popularized the *sannong* issue. These left-leaning scholar-activists reintroduced to the public the RRM legacy in the early 2000s and were referred to as the "New Rural Reconstruction Movement" (Pan 2012; Day 2008, 2013a, 2013b; Jacka 2013; Yan and Chen 2013; Yeh, O'Brien, and Ye 2013). The NRRM's work across the country touches upon various issues including ecological agriculture, civilian education, farmers' cooperative[5] facilitation, civil rights protection of migrant workers, rural sustainable development in general, and so on (see Table 9.1). Mainly engendered from concerned intellectuals, the NRRM's sentiments resemble the ideas put forward by the RRM in the 1920s. These ideas include critiques of elite culture and knowledge, focusing on peasant status and subjectivity, connecting elite intellectuals with the rural masses, reconstituting rural–urban relations, experiments with the rural education reforms and improving rural health care conditions, etc. (Pan and Du 2011a). In response to more contemporary concerns, mitigating ecological crisis has been added as an important task for the NRRM to address (Pan and Du 2011a). Day (2008, 50) noted that, "as a critique of developmentalism and the economic mode of analysis," the NRRM "turns to culture and cooperative relations as vital to the reorganization of rural social life." It is a response to the dominant neo-liberal logics of marketization, seeking alternative modes to revive the rural economy, culture, society, and

Table 9.1 The NRRM projects and experiments (adapted from the promotional video of the NRRM and information collected from interviews)

Categories	Projects	Objectives	Year initiated
Research	Rural Research by University Students to Support Rural People	**Serving the peasants*** and striving for dreams	2001
	Rural Reconstruction Centre of China at Renmin University	People's livelihood centred, collaboration and cooperation, multicultural base	2005
	Institute of Rural Reconstruction at Southwest University	Sustainable development and rural reconstruction, "beautiful countryside" research	2012
	College of Strait Rural Development at Fujian Agriculture and Forestry University	Farmers collaboration, financial support for farmers, ecological agriculture and "civil agriculture" development, promotion of community colleges	2013
Integrated Experimental Zones	Puhan community, Yongji, Shanxi Province	Integrated Community Construction for **Sustainable Rural Development**	1998
	Zhaicheng county, Dingzhou, Hebei province	Economic, Cultural, Education and Medical Treatment Integrated Community Development Project	2003
	Lankao, Henan province	Farmers collaboration and rural-urban cooperation experiment, multi-stakeholder involvement: intellectuals, college students, grassroots, local government and urbanites	2003
	Shunping, Hebei province		2004
	Wujin, Jiangsu province	Experimenting with ecological civilization, rural culture in **new countryside construction**	2010
	Gangli, Yushi county, Henan province		2011

Category	Location	Description	Year
Farmers' cooperatives	Nanmazhuang, Lankao, Henan province	The first ecological village in Central China	2003
	Nantang, Fuyang, Anhui Province		2003
	Sancha, Fangxian County, Shiyan, Hubei Province		2003
	Jiangzhuang, Yutai County, Shandong Province		2004
	Shangping, Yong'an, Fujian Province		2005
Civilian education (community colleges)	James Yen Rural Reconstruction College in Dingzhou, Hebei	Ecological agriculture and environmentally friendly countryside, the first free civilian education school in the countryside in the twenty-first century, promoting local rural knowledge and **scientific development**	2003–2007
	Shiwu Community College, Danzhou city, Hainan province	The first rural community college in China, exploring a potential path for rural adult education	2006
	Fuqian Rural Reconstruction Center, Anxi county, Fujian province	Promoting sustainable and integrated rural community development, cooperative economy and sustainable agriculture	2009
	Peitian Community College, Liancheng county, Fujian province	Protecting traditional villages and preserving rural culture	2011
	Tingtang Community College, Putian, Fujian Province		2011
	Dahu Community College, Nanchang, Jiangxi province		2012
	Jiaocun County Rural Experiment, Lingbao, Henan	Promoting traditional moralities and rural culture, rediscovering indigenous knowledge	2012
CSA farms	Beijing Little Donkey Farm	An integrated platform for farming, citizen education, research, CSA promotion and personnel training	2008

continued

Table 9.1 Continued

Categories	Projects	Objectives	Year initiated
	Big Buffalo Farm	Promoting traditional farming knowledge, CSA model, Participatory Guarantee System,[a] **harmonious urban–rural development**; rebuilding the trust between urban and rural communities	2011
	Little Donkey Liulin Community Farm, Beijing	Creating an agricultural community shared mutually by urbanites and farmers in suburban area	2012
	Guxiang Farm, Fuzhou, Fujian province	Awakening people's love for their hometown	2012
	Shared Harvest Farm, Beijing	Strengthening the connections between urban citizens and farmers, securing ecological farmers' income, improving food safety conditions	2012
Migrant worker centres	Beijing Migrant Workers' Home	Improving living conditions of migrant workers, defending migrant workers' rights (New Migrant Worker Arts Group, Cultural Festival)	2002
	Green Ground Migrant Workers' Home, Xiamen, Fujian province	Education and development of migrant workers in emerging industries	2007
Devotion to homeland network	Various "devotion to homeland" projects	Promotion of indigenous knowledge, protection of rural culture, rural reconstruction personnel training, rediscovering the value and **beauty of the countryside**	2012
Other initiatives	Ecological Architecture Studio	Intellectuals working with peasants to promote ecological architecture in rural area and earthquake-stricken area	2004
	China Office of Global Peace Women	Promoting women's rights and grassroots women communication	2005
	Beijing Green Ground Union	The first farmers' green food production association in China	2006
	Beijing Green Ground Cooperative	The first consumer cooperative in China: healthy consumption and fair trade	2006

Aoxiang Society of Beijing Forestry University	Campus farming, sustainable campus experiment (National College Students Campus Farming Symposium)	2008
Chongqing Green Ground Dapinghuo Community Restaurant	Rallying with local small farmers with the power of consumption: environmentally friendly farms, healthy life and community	2012
Slow Village Project	Supporting youth return to their home villages, revitalizing rural economy and culture, promoting "**targeted poverty alleviation**"	2017

Notes

* Bold words indicate the use of mainstream discourses.

a According to IFOAM, Participatory Guarantee Systems (PGS) are "locally focused quality assurance systems … that certify producers based on active participation of stakeholders and are built on a foundation of trust, social networks and knowledge exchange." For more detail please refer to: www.ifoam.org/fr/value-chain/participatory-guarantee-systems-pgs (accessed September 16, 2014).

peasants' subjectivity (Pan and Du 2011a; Yeh, O'Brien, and Ye 2013). From a developmental perspective, it represents civil society's attempts to seek a self-organizational approach to rural development.

Table 9.1 demonstrates that most of the projects of the NRRM were launched after 2003. The establishment in 2003 of the James Yen Rural Reconstruction Institute in Dingzhou, Hebei province, was emblematic of the revival of the RRM from the 1930s. Dingzhou (previously called Dingxian), where James Yen's rural reconstruction efforts took place from 1926 to 1937, was regarded as one of the cradles of the RRM (see Day 2013b). The NRRM team not only founded the Rural Reconstruction Institute but also revived Dingzhou as a symbolic experimental zone for rural reconstruction. This unofficial "first year of the NRRM" was also marked by the establishment of several farmers' cooperatives in Henan, Anhui, and Hubei, soon followed by a few other projects. A few earlier projects like the Beijing Migrant Workers' Home and a few earlier farmers' cooperatives soon joined the NRRM team. It is interesting to note that all the CSA farms[6] were started after 2008. Although a few more community colleges (*shequ daxue*) were established after 2008 to fuel their long-term goal of civilian education, CSA farms became the star projects of the NRRM and attracted the most public attention. This raises the question of why this is the case, and how the NRRM's involvement in AFNs, particularly CSA farms, affected its development in China.

Since 2012, the various initiatives of the NRRM have launched the "Devotion to Homeland" (*ai guxiang*) campaign to re-pitch the advocacy of rural reconstruction, that is, to form an alliance between university students, "returned youth" (people who returned to their hometowns from cities), migrant workers, and artists and to explore the potential of this alliance to facilitate the sustainable rural development. The core objectives of this campaign are to rediscover and revalorize the countryside through awakening the deep emotional connections with the countryside deeply rooted in Chinese culture and transforming that nostalgic complex into a new force of the NRRM. Most recently, in 2017, the NRRM began to collaborate with Slow Food International to promote the establishment of slow villages in China. This new initiative marks a big step forward of the NRR team to collaborate with the private business sector such as the catering industry.

Challenges facing the New Rural Reconstruction Movement

As previous chapters of the book explained, under the authoritarian state in China (or semi-authoritarian state; see Ho and Edmonds 2007), democratic movements have been mainly repressed or exiled, and this reflects a weakness of Chinese civil society. Ho and Edmonds (2008) coin the term "embedded activism" to describe the situation of advocacy in China. Yet, social movements like the NRRM still enjoy some freedom, as long as they do not overstep the state's power realm of repression. Therefore, on the one hand, collective action groups need to impose self-censorship and de-politicize their activities to stay away from the state's radar of political sensitivity. Their strategies to deal with the state have to be non-confrontational in order to maintain their legitimacy. On the other hand, collective action groups have developed informal personal connections with the state, via retired officials or state-run organizations, which gives them a certain range of freedom to facilitate their activist goals.

Although the NRRM is not a movement that pursues political power, it is inevitably political given that social forces in this process are not under the complete control of the state but rather under the leadership of a group of intellectuals and NGOs. Interacting with the state at various levels is hard to avoid. In fact, despite the fact that the first wave of RRM in the 1920s led by Liang Shuming attempted to exclude the state from community projects by working directly with peasants and keeping independent from the state, Liang's experiment of civilian education in Zouping, Shandong in the 1930s could have never happened without the assistance of the local warlord (Thøgersen 2009). Thøgersen (2009, 29) argues that "a general uneasiness about state actors manipulating and dictating rural communities and a growing feeling that classic CCP governing mechanisms are unable to solve the problem of community building dominates the present discourse on rural reconstruction." Thus, the NRRM is constantly seeking an alternative to the top-down state-led rural development approach (Day and Hale 2007) or, in Pan and Du's (2011a) words, an alternative to "the modern dream."

"Anti-modern" sentiment and interventions of the pro-modernization state

Social movements in the "restrictive political environment in which various socio-economic and cultural changes are taking place" (Ho and Edmonds 2007: 2) face various challenges from the state and society itself. As a social movement that aims to counter the modernization ideologies (Pan and Du 2011a), the most fundamental challenge the NRRM faces rests on their will to explore an alternative rural development approach, in contrast to the state's approach of urbanization, commodification and marketization of the rural. Wen Tiejun, the most renowned advocate of the NRRM, noted that the formation and expansion of capital in the history of colonization is the cause for environmental deterioration of the colony and reflects the theory of "'modernization' that we take for granted today" (Wen 2007, 13). Wen's critiques of ideas of modernity such as privatization, commodification, marketization, globalization, liberalization, and democratization largely resemble the neo-liberalization trend that has been promoted in the developing world. Hence, the NRRM challenges the hegemonic development thinking that has been guiding the development of China since the economic reform started in 1978 (Day 2013a).

The "reflection," a term used by Wen (2007), lays the foundation of the NRRM's social experiments at a local, regional, small-scale level with a focus on empowering peasants amid the wave of globalization, marketization, and urbanization. The NRRM team believes that the critical reflections on, or the deconstruction of, modernity and their social experiments in rural China forge a new understanding of modernity—"anti-modern modernity" (see Wang 1998)—which runs through the thinking of the two waves of rural reconstruction (i.e. the RRM and the NRRM).

We argue that the closing of the James Yen Rural Reconstruction Institute at Dingxian county, Hebei province in April 2007 reflected some operational and ideological conflicts between the NRRM and the state. For three years and nine months the James Yen's revived civilian education programs conveyed not only a bold declaration of alternative development but also the non-compliance of the NRRM's activities with the state's trajectories. The institute was shut down by (in the name of) the Bureau of Education in Dingzhou city, as the Bureau claimed the operation of the institute illegal. An advocate of the NRRM noted in our interview that their eco-architecture model was also claimed illegal by the local government. The village party secretary of Zhaicheng village, where the institute was located, emphasized that, although the institute received enormous attention and support from society, it would not last without a supportive "political environment" (Weng 2008), an environment in which the state endorses and encourages the operation of alternative developmental programs. The opposition from the local state partly came from the values of the institute—promoting the solidarity of peasants. The trainees, mainly villagers, were required to exclaim a slogan every day before lectures: "be the master of homeland by changing ourselves, building a new countryside by uniting ourselves" (Xu 2011). This could be interpreted as a sign of revolution with political implications. Instead of focusing on "modernizing the village" (economic development), as the local government wanted, the institute aimed to establish peasant solidarity and subjectivity. This offered no economic benefit to sustain local government's support of the once promising project.

The state's intervention in the programs of the NRRM is also reflective of its unwillingness to allow large-scale peasants' alliances owing to the concerns about social stability.[7] As Thøgersen (2009: 30) noted in his analysis, "farmers are encouraged to solve their own problems through intra-village cooperation, but they are not supposed to organize across administrative borders … they depend on the goodwill of state actors." The party secretary of Zhaicheng village explained that the reason behind the shutdown of the James Yen Rural Reconstruction Institute was that it expanded its educational program to include farmers' cooperative training for people from other parts of the country, which exceeded its original sphere of operations. The overwhelming media coverage on the "alternativeness" and "otherness" of the project also put pressure on local authorities.

According to Yu (2011), more than 1,000 peasants, village leaders, and volunteers visited Zhaicheng for training and other activities. With all of these political issues surrounding the institute, it couldn't get official registration after 2005.

Social disjunction

We see another critical reason for the NRRM's difficult acceptance into society and that is "social disjunction," a challenging situation where civilians do not see the need for or the benefits of the NRRM. As one villager from Zhaicheng village commented on the James Yen Rural Reconstruction Institute's work,

> James Yen taught us reading and promoted new crop varieties at a time when we needed (the service) but couldn't get it. Times have changed and the condition is no longer the same, but they (the NRRM) have the same ideas.
>
> (Weng 2008)

This disconnection between the NRRM and people's needs is critical.

The social disjunction is specifically reflected in the NRRM's work in ecological agriculture promotion. The early attempt at organizing local peasants to do ecological farming was arduous. In 2003, when the program started, only four households out of more than 1,000 from Zhaicheng village joined the ecological farming experiment after the institute proposed to subsidize 200 CNY for each *mu*[8] (US$199.4 per acre) of farmland. Local villagers ridiculed ecological farming methods that abandoned synthetic fertilizer and chemical pesticides and herbicides and worried that the pests from the ecological farming plots would cause damage to their crops (Lin H. 2008). The practice of ecological farming was labelled as an unrealistic plan brought by a group of idealists who were not familiar with practical farming in the area. In our interview with one of the leading advocates of NRR, he narrated the story of a donkey, which later became the symbol of the leading CSA in China (Little Donkey Farm).

> My ten years' work in NRR taught me a lesson—ecological problems of rural China are not at all technical problems, they are closely related to the broad social economic background. Let me tell you a story about a donkey.... We treated it as a symbol of ecological agriculture and a challenge to petroleum based agriculture because it is a symbol for using animal labour instead of fossil fuels. However, when they started to raise the donkey in Hebei province in 2005, we received objections from villagers who wished they could bring more modern agricultural technology and believed a donkey is a symbol of backwardness. After we brought the donkey to Beijing in 2008, we thought

it would be a good help to ecological farming. However, as there was no place to cut the donkey's hoof and no old peasants who knew how to harness the donkey, it became useless and later only a symbol. The embarrassment over the donkey shows that if you want to change the current system, you have to change it entirely and fundamentally.

(Interview with a key figure of the NRRM in Chongqing, China, May 6, 2012)

The "donkey dilemma" made the NRRM community realize that the challenge of implementing ecological agriculture in China goes beyond the farm. It is an integral challenge, facing not only technical challenges in farming but also challenges within society. Peasants refused to plow with donkeys because they had been told through other education that the donkey-associated traditional farming system was backward and needed to be replaced by modern and high-tech farming (Schneider 2015). Petroleum-based farming has become an agricultural norm and a symbol of "good agriculture" that persists in the mind of Chinese peasants, even in those peasants who do not have access to the modern technologies.

Not only were there misunderstandings with local villagers but urbanites also could not understand the NRRM advocates because of the perceived notion, or "common sense," that professors in "ivory towers" do not work directly with lowly peasants in the field. However, this perception was strongly challenged by a news story in 2006—"professors selling rice." He Huili, an associate professor at China Agriculture University and an NRRM advocate who facilitated the farmers' cooperative in Nanmazhuang village, Henan province, attempted to sell in Beijing hazard-free[9]

Figure 9.1 The namesake of Little Donkey Farm in Beijing (that passed away in 2018).

rice produced by this cooperative. Despite the endorsement of Wen Tiejun, the sale did not go smoothly at first. The inconsistency between the respected title "professor" and the mundane "rice seller" created material for media coverage, often satiric, and opened debates. He Huili and Wen Tiejun were criticized for not obeying the principles of market economy, and for their "anti-market" behaviour (Tong 2006; Wang 2006). Our interviews revealed that they were cheated and lost their first sale when their rice was delivered but not paid for.

The disconnect between the NRRM's ideal image of agriculture and the Chinese society's stereotyped image and solidified longing for modernity reveal a severe challenge to intellectual-led rural developmental projects. When taking the state interventions into consideration, the obstacle becomes even more difficult to overcome. Nevertheless, the rapid change occurring in Chinese society offered an opportunity to the NRRM to obtain the support of peasants and urbanites, and in certain cases even the state. This chapter analyzes how the NRRM has coped with these challenges by strategically using avenues heavily loaded with mainstream development ideas that accord with the state's will. We argue here that AFNs, especially CSA farms, were pushed by the society's food scares to the frontline of the NRRM. AFNs became an effective tool for the NRRM to carry out their alternative experiments of rural development while, at the same time, achieving public support.

Coping with the state and the social context by promoting alternative food networks

"Whoever understands the times is a great man." This Chinese saying, embraced as a motto by many Chinese, is used too easily to blame rigid structural forces as a convenient excuse for failure, while also acknowledging the detrimental impacts of socio-political conditions. In recognizing the state's rural development agenda as being largely modernization-oriented (with urbanization and industrialization embedded within this), the NRRM team has had to use adaptive strategies to cope with this structural arrangement. They developed appropriate tactics to capture, manipulate, and apply relevant state initiatives to achieve their own goals. This chapter illustrates these approaches from two dimensions. One is how the NRRM has used mainstream discourses to politically justify and promote their alternative initiatives; the other is how they seek a harmonious relationship with the state by seeking common ground.

Using mainstream discourses

In justifying the NRRM's alternative logic, Wen (2007) argues that Western modernization based on 300 years of colonization cannot be replicated in China and, thus, is not a "scientific" concept that can guide the development of China. Modernization has been deconstructed in many different ways in development studies (see Nederveen Pieterse 2010) but there has not been an examination of the concept questioning its "scientificity." This odd but novel angle of deconstructing modernization makes much more sense when linked with the political slogan "scientific approaches to development." The Chinese Communist Party in Hu Jintao's era summarized the guiding developmental thinking of the country as "scientific approaches to development" (*kexue fazhan guan*) in 2007, giving rich meaning and also a pivotal position to the term *kexue* (scientific) in development policies, requiring government at every level to make policies based on whether a policy is "scientific." It was later written into the constitution of the CCP as a guiding ideology of the CCP's work. Thus, examining the scientificity of Western modernization becomes relevant and significant in the Chinese political context.

Attempts to stress the political relevance of the NRRM were also represented by how He Xuefeng, another leading advocate of the NRRM, reinterpreted the term "socialist" in the NSCC campaign to include more soft values promoted by the NRRM. He X. (2007: 30) argues that

> the word "socialist" in the new socialist countryside is not an empty word but one of great significance. *Non-market factors in the villages may be mobilized* for social, cultural, and organizational construction in the countryside that may result in a large increase of *non-economic benefits for peasants*. In addition to economic income, peasants may also obtain cultural and social benefits as well as benefits in terms of *decency* and *dignity*.

> (Emphasis added)

This is a strong rebuttal to the tendency of over-simplification of *sannong* as an economic issue in mainstream developmental approaches.

Recognizing the power of discourse in providing a solid political base for their initiatives, the NRRM team adopts mainstream terms to promote their alternative and community-based experiments. Table 9.1, within which the objectives and descriptions of their projects were directly translated from their promotional documents, highlights some of their manipulations of mainstream words (phrases in bold). For example, the Wujin experimental zone claims to "experiment with ecological civilization, rural culture and new countryside construction." "Ecological civilization" became a buzzword in 2012 after it was included at the Eighteenth People's Congress as one of the five major developmental tasks of China.[10] Although official policies guided by this concept have been limited and the state still hesitates in promoting ecological agriculture on a large scale (Scott et al. 2014), the NRRM experiment in Wujin of establishing ecological farms realized the importance of the discourse. They even use "new countryside construction" rather than the NRRM to obtain a certain degree of political relevance. Other examples are the use of political buzzwords like "promoting scientific development" and "harmonious urban-rural development."

Seeking a harmonious relationship

Using mainstream discourses has been one strategy of the NRRM to seek a harmonious (non-confrontational)[11] relationship with the state and the mainstream. Indeed, the most solid common ground between the NRRM and the state is their shared goal of tackling the *sannong* issue. In an interview with *Southern Rural Daily*, Wen Tiejun explained his concern about the radicalness of the term "reconstruction" (*Southern Rural Daily* 2005, translated from Chinese).

> The international, commonly used term for the rural construction movement is "rural reconstruction" [*xiangcun chongjian*], but I am a moderate person and I don't want others to misunderstand or misinterpret our works, so we call it "new rural construction" [*xin xiangcun jianshe*, avoiding the term *chongjian*], but the English translation is still "rural reconstruction."

"Reconstruction" (*chongjian*) implies a process of deconstruction, revolution and structural changes while "construction" (*jianshe*) is a plain term that has no strong connotation. As a way to maintain political sensitivity, translating NRR as "new rural construction" is tricky but clever. The NRRM team has emphasized that their major approach is "amelioration" rather than "revolution" (Pan 2012), again implying its willingness to maintain a moderate manner.

In the 2012 annual CSA symposium, the Rural Reconstruction Center at China Renmin University, together with several other academic institutions, called upon CSA farmers to forge a national "Ecological Agriculture Cooperation Network" (*shengtai nongye huzhu wangluo*). This organization aims to form an internal monitoring mechanism and facilitate information flow among its CSA members. However, collective actions are often perceived as a threat to social stability and state authority in China. To address this risk, Wen Tiejun, the convener, expressed the role of the network in a modest manner.

> What we are currently doing is merely a continuation of the previous exploration of an alternative developmental path that happened almost one hundred years ago. We don't actually cause harm to any interest group. We also don't cause negative impacts on our current policies.... It [organizing an ecological agriculture cooperation network] is only a small activity that accords with our big ecological civilization agenda.
> (Transcribed and translated from Wen Tiejun's remarks at the launching of the National Ecological Agriculture Cooperation Network, November 30, 2012)

Another recent key change we observed is the rename of the NRRM in Chinese, for political reasons. Activists we talked to in 2016 and 2017 started to refer to the NRRM as "new rural reconstruction network," replacing the term "movement" (*yundong*) with "network" (*wangluo*). This clearly mirrors that the NRRM seeks to shun the politically sensitiveness of "social movement" in China. One of the key figures of the NRRM told us:

> Politically in China the term "movement" isn't the terminology we use, it is kind of transformation or change, system change ... network is a neutral term. And it is really a network. No mandatory positioning, no leader nor institutional setting for the Rural Reconstruction Network—no headquarter, no leadership, no centrally controlled fundraising.... We need to be very strategic in China. It is quite risky to have an institutional structure. The decentralized structure and lack of attachment to a specific institution keeps us more resilient.
>
> (Interview with a key figure of the NRRM in Beijing,
> August 19, 2017)

These cases are just a small piece of the NRRM's efforts in forging a harmonious relationship with the state and the private sector. Wen Tiejun, the leading figure of the movement himself, has in fact been viewed as a quasi-official who has an entangled relationship with the state and whose proposals can shape policies (Day 2008). This double role provides both challenges and opportunities for the movement's status in the political realm. The complexity of their relationship with the state requires them on the one hand to maintain a relatively moderate profile while on the other hand to lead the social activities looking at alternative development approaches. How this contradictory role will shape their advocacy is yet to be seen.

Promoting alternative food networks[12]

While the NRRM faces various challenges from the state and society, its initiatives with AFNs have demonstrated its social and political relevance as well as its vitality. This fascinating process shows how a social movement could magnify its impacts by responding in a timely way to social changes. In the NRRM case, one of the key social changes in the past few years was the loss of citizens' trust in food safety due to various food safety scandals (see Pei et al. 2011; Wang et al. 2015). The food safety issue transformed the way that the ecological farming work and sustainable food initiatives of the NRRM were viewed by the public.

The compelling story of the NRRM's involvement in AFNs dates back to 2003, when they facilitated hazard-free certified rice production in Nanmazhuang, Lankao city in Henan province. The original production model of the farmers' cooperative was not an effective approach to acquiring customers' trust. The "cynical" society, to use an interviewee's word, was highly skeptical of the rice, which is more expensive due to its low yield, despite it being endorsed by renowned professors. However, the situation changed completely in 2008, one year after the closure of James Yen Rural Reconstruction Institute (which ironically failed to promote ecological farming in Zhaicheng village). The melamine scandal, coupled with other food safety scandals, created anxiety in the general public. People suddenly became extremely passionate about searching for safe and healthy food and organic food and other types of ecologically produced food rose in popularity among Chinese consumers. It was at that time that the first CSA project of the NRRM—Little Donkey Farm—took off in the suburb of Haidian district in Beijing. Although both Little Donkey Farm and the Nanmazhuang Farmers' Cooperative followed ecological farming approaches and were directed and endorsed by intellectuals, they had quite different public receptions. One interviewee, who participated in the two projects, acknowledged the sharp contrast in our interview.

> In 2005, He Huili's efforts to promote Nanmazhuang rice got much media attention, but roughly only one third of the media coverage was sympathetic, while one third had no position and one third was satirical (ridiculing her). In 2008, Shi Yan—also an educated woman promoting ecological agriculture—got lots of media attention in establishing Little Donkey Farm, and 99% of coverage was positive. What made the difference? It was the rise of food safety concerns since 2008.
>
> (Interview with a key figure of the NRRM in Chongqing, China, May 6, 2012 (translated from Chinese))

Little Donkey Farm soon became the leading CSA farm in China. Our visits to various CSA farms in China found that Little Donkey Farm was viewed by many as a model and source of inspiration.[13] The proliferation of CSAs across the country led to an annual national CSA symposium being started in 2010. The influence of Little Donkey Farm later went beyond CSAs and ecological agriculture as it also facilitated some of the earliest farmers' markets and buying clubs in China. They have close connections with the first ecological farmers' market—the Beijing Farmer's Market—and related farmers' markets in Shanghai, Xi'an, Guangzhou, and Tianjin, as well as Shanghai *Caituan* Buying Club and Huilongguan Buying Club in Beijing. The farm is embraced by urbanites not only for its ecologically produced vegetables but also for its rental plots,[14] which allow urbanites to rent plots to grow vegetables.

Although the farm became a symbol of a civil society initiative in solving food safety problems, it is necessary to recognize the state's support in its establishment and development. According to our interviews,[15] the government of Haidian district facilitated their access to farmland and subsidized the infrastructure construction.[16] Little Donkey Farm is officially titled the "integrated production, learning and research base co-founded by Haidian district government and China Renmin University." The farm was endorsed by the local agriculture and forestry bureau.[17] As another example of state support for this style of project, the Big Buffalo Farm was established in 2011 as a collaborative project between China Renmin University and the government of Wujin district in Changzhou city.[18] The collaboration between civil society initiatives and the state depends upon their common ground in agriculture development, especially the state's policy in promoting multifunctional agriculture (recreational agriculture and innovative agriculture[19]). This forges a base for the NRRM and the state to work together.

The growing public anxiety over food safety unveiled a whole new horizon for the NRRM. AFNs, especially CSAs, that won the hearts of both society and the state, became powerful tools for the promotion and implementation of the NRRM's alternative developmental ideas. AFNs help to concretize the movement's idealistic values and reconnect it with the demands of society. Food thus emerged as a promising hope for the NRRM amid the cynical views and complicated expectations of society and the state. The NRRM team successfully captured the opportune moment to establish several other CSA farms, including the Big Buffalo Farm in Changzhou, Jiangsu province, the Little Donkey Liulin Farm in Beijing, and the Guxiang Farm in Fuzhou, Fujian province (see Table 9.1). It takes full advantage of Little Donkey Farm as a platform for environmental education targeting customers, visitors, volunteers, and farmers who are interested in the CSA model. CSA farms became a much more attractive place for civilian education compared to the rural reconstruction institutes established by the NRRM team. It offers a chance for urbanites to take a closer look at the value of agriculture and "the rural" as a whole, which underpins the entire NRRM.

Despite the success of CSA farms, the practices of these CSA farms embodied tensions and conflicts, some of which challenged the NRRM's values of social justice and reviving peasants' subjectivity. Instead of being integrated into the decision-making process from the beginning of the operation, peasants were mainly hired farm workers who devoted their labour to the farm.[20] Throughout our fieldwork, under the discourse context of *suzhi* (quality), within which peasants are normally perceived as having low "quality" (see Schneider 2015), peasants were not trusted by consumers and some initiators of AFNs. CSA farms under the NRRM umbrella are mainly ventures of intellectuals with strong entrepreneurship within which peasants are largely marginalized (see also Schumilas 2014: 167–168).

Policy implications of the New Rural Reconstruction Movement

Gaining the support of ordinary consumers did not automatically translate into policy changes, and so the interactions between the NRRM and the state have another facet, which is the NRRM seeking policy changes. Thøgersen (2009: 26) argues that the state is willing to cooperate with academic advocates because it believes they can generate new ideas and perspectives. Indeed, grassroots and civil society initiatives, which always break current rules, can be developed into future policies in China's distinctive policy innovation process (Heilmann 2008a, 2008b).

An example is the seniors' organization experiments led by the NRRM (Wang 2009). He Xuefeng (2007) emphasized that the countryside could contribute to social stability by becoming an "emotional and meaningful home" for migrant workers in cities. The NRRM enriches peasants' social, cultural, and spiritual life by facilitating self-organizations (such as seniors' associations and cultural performance troupes) (see Qiu 2007) that make the countryside a "meaningful" space and thus "a homeland to which one could return" (He 2007: 36). In this sense, the core of rural construction should be "social and cultural construction" that enables a way of life with "low consumption and high benefit" in contrast to the mainstream policy based on a market economy to induce high rural consumption.

The policy implications of the NRRM were summarized by Wen Tiejun in his new interpretation of *sannong* based on his deconstruction of modernization (Wen and Sun 2012). The rural problems in three dimensions (agriculture, villages, peasants) have long been a set of challenges, including maintaining "the growth of agriculture productivity, the development of villages and the increase of peasants' income." These three developmentalist goals have been guiding China's policy making and developmental agendas for the rural countryside. The NSCC, for example, is a state-led campaign that was designed to meet these three goals. On criticizing the oversimplified economic orientation and the absence of sustainability and social justice concerns, Wen Tiejun proposed a new interpretation of *sannong*, specifically in reference to the three rural development goals. In contrast to the old *sannong*, he interprets the new one as "the protection of peasants' rights, the sustainability and stability of the countryside, and the ecologicalization and safety of agriculture" (Wen and Sun 2012, 11). This new interpretation saves *sannong* from an economic-focused oversimplification and recovers its multiple connotations. It highlights the role of "the rural" in social stability and justice as well as ecological sustainability, which were downplayed or even ignored in previous policies tackling the *sannong* challenges.

Discussion and conclusion

In analyzing the dynamic state–society relationship in China, Saich (2000) argues that Chinese social organizations often live in a symbiotic relationship with the state. On the one hand, NGOs' increasingly important roles in environmental protection and solving social problems are recognized by the state and the state is increasingly dependent upon NGOs to provide social services; on the other hand, NGOs need the state's support, or at least passive acceptance, to fulfill their objectives. The story of the NRRM is an example of not only the symbiotic relationship itself but also how this relationship is forged. It is a process within which the state adopts the wisdom of intellectuals, and the social organization piggy backs on the state's authority. These mutual interactions in the NRRM case are reflected by the policy implications of the NRRM as well as by the NRRM taking advantage of the mainstream developmental discourses and policies. NRRM experiments in local contexts provide ample experiences in solving social and environmental problems that the state can take into consideration. The critiques of the NSCC being too economically oriented could be largely addressed by the NRRM's work in civilian education, peasants' social organizations, cultural group facilitation and ecological agriculture promotion. In addition, the NRRM also borrows political discourses proposed by the state, such as "scientific developmental approaches," "ecological civilization," and "harmonious urban–rural development," to justify its political rationality and to seek further support from the authorities. Its efforts in coping with the state also demonstrate the construction of a harmonious relationship with the state created by avoiding aggressive and revolutionary terms and emphasizing it being a movement of "amelioration" rather than "revolution."

The interactions of the NRRM and the state exemplify some adaptations and accommodations of institutional arrangements and policies in both directions: the NRRM inspired the rural development agendas of the state, at least in terms of bringing the significance of rural issues to the forefront of the policy and institutional realm. As the Chinese government is transforming its rural developmental agendas toward more integrated ones, the experiences of the NRRM in addressing social, cultural, and sustainability challenges offer valuable resources. Meanwhile, the NRRM found itself striving to adapt to the state's interests. This bilateral accommodation between the state and the NRRM, we argue, illustrates an effective scaling-up strategy of civil society initiatives in the entrenched state-led realm of rural development and thus has significant implications for the scaling-up of community-level, local, and small-scale endeavours in various fields.

Moreover, the complexity of socio-political contexts, in which the NRRM is pushing forward its alternative and often anti-modern developmental agendas, is represented not only by the interactions with the state but also the interactions with society. Here, "society" embodies the peasants with whom they are directly working and the urbanites and other intellectuals who hold contradictory perspectives about their work. Positioned in the middle of the state–society dichotomy, the NRRM is highly dependent upon society's support to make policy impacts, yet large social support and participation might be perceived by the state as a threat to social stability. The early attempts of the NRRM in promoting ecological agriculture, however, encountered very limited support from peasants but misunderstandings and critiques from urbanites. The situation changed after the food safety crisis in 2008, when the NRRM started a CSA project—Little Donkey Farm—in suburban Beijing. The CSA project was not just a response to the crisis, however, since their promotion of ecological agriculture had begun much earlier. The synergies between their CSA work and the rising food safety anxiety made AFNs effective tools for promoting ecological agriculture and organizing peasants. It also enabled the general public to rediscover the value of the countryside which had long been marginalized in the development of the market economy.

The involvement of the NRRM in AFNs development in China exemplified an interesting phenomenon in which AFNs functioned not only as a tool for transforming the food system but also as an accelerator for achieving broader rural development goals. AFNs, while aiming to foster rural development and tackle the food safety problem, opened a convenient and promising space to develop civil society capacity, as the AFNs were less confrontational and politically sensitive than other society-driven initiatives. It enabled the NRRM to demonstrate to the state its potential in solving urgent social problems in a non-confrontational manner. More importantly, it became a powerful instrument to concretize the often-criticized romantic and idealistic values of NRR and reconnect them with the demands of society. In this sense, AFNs helped the NRRM to build a solid political and social foundation.

As a matter of fact, the relationship between AFNs and rural development is indeed reciprocal. Working with food activists and non-governmental organizations, the NRRM was able to bring CSA farms, ecological farms, and other AFNs together under one umbrella given their common alternative rural development agenda. On the one hand, the NRRM has been using AFNs to tap into environmental and ecological agendas of the state and cope with societal demands for safe and healthy food; on the other hand, the NRRM has created a platform for the convergence and scaling-up of locally specific and fragmented AFNs (see Si and Scott 2016).

What is more promising for the NRRM is that there are continuing opportunities emerging for them to connect with state developmental agendas and the expectations of the masses. For example, their endeavours in AFNs and ecological agriculture fit with the state's adjustments of the developmental orientation of the agriculture sector. "Multifunctional agriculture," proposed in 2006, "ecological civilization," proposed in 2007 (see Wen et al. 2012), and "two orientations of agriculture" (resource-saving and environmentally friendly agriculture), proposed by the central government, indicate a gradual transformation of the government's policies in agriculture development. In May 2015, the Chinese government issued the National Sustainable Agriculture Development Plan (2015–2030), which specified key tasks for sustainable agriculture development. These policy changes all accord with what the NRRM has been working on and thus, can provide opportunities for the NRRM in the future. Another emerging opportunity is the growing number of migrants, returning to their rural home from urban areas in China (see Démurger and Xu 2011, Fan 2008), who can become a powerhouse for the NRRM's initiatives. Our interviews with CSA farmers in Beijing, Shanghai, and Chengdu clearly revealed that many CSA farms were established by people who returned from the city with knowledge and ideas that they gained from their urban experience.

Although this chapter identifies the strategies of the NRRM in coping with the demands of the state and society, it is still unclear how these strategies shape their initiatives. What the NRRM will be like in the future and whether AFNs will lead to a food-based NRRM is worthy of further study. Despite the remarkable achievements of the NRRM and the emerging opportunities with the adjustments of state policies and the new trends in society, whether the NRRM will become the prelude of a more fundamental paradigm shift that leads to a "new rural development paradigm" (see Goodman 2004; Watts et al. 2005; Renting, Marsden, and Banks 2003; Marsden, Banks, and Bristow 2000; Tovey 2009) in China is still unknown.

The NRRM's ideas and practices also deserve further critical scrutiny, in that it is not a coherent movement without tensions and contradictions. Although the crisis facing rural China is akin to that in 80 years ago, it is problematic to transplant the strategies from the RRM directly to contemporary China. The capital's infiltration of peasants' rural life, the mainstream urban-inclined rural development orientation, and the public's heightening food safety anxiety all demand novel strategies and tactics from the NRRM. In examining the farmers' cooperatives of the NRRM, Hale (2013) argued that the NRRM unintentionally introduced "capitalist relations" to their cooperatives practices, which contradicts their anti-modern sentiment. Conflicts between the capitalist form of value and the NRRM's alternative value have been shaping the NRRM's projects. Going against the urbanization and industrialization process bestows a strong alternative feature on the NRRM but also constrains the movement's scale and influences in the Chinese context. In addition, as Yan and Chen (2013) pointed out, the NRRM assumes the state as "the public power that rises above various interests" (p. 976) and requests that the state restrain capital's influence on rural China. This dependence on state power to achieve social reform goals indeed reflects the symbiotic facet of the relationship between the NRRM and the state. However, it also fundamentally challenges the NRRM's degree of freedom when exercising its alternative values.

Notes

1 "Civilian education" (or mass education) promoted by the NRRM is in contrast with the official educational system which is referred to as "elite education," an "appendage" of politics that aims to train people to become elites, leave the countryside, and serve the dominant hierarchy. Rather, "civilian education" teaches villagers who are lack of educational opportunities to be literate and obtain livelihood strategies, knowledge, and techniques that they can use in the countryside.

2 *Sannong wenti* was also phrased as "the rural problem in three dimensions: village communities, agriculture and the peasantry" (Pan and Du 2011a: 454).

3 Reconstructing the subjectivity of Chinese peasants has been a critical element in the NRRM. It refers to the NRRM's belief that peasants should not be the passive recipient of social transformations but should and can be the subject that proactively participates in and creates social transformations (see Pan and Du 2011a; He, Cheng, and Zong 2014).

4 China Fortune (Xu 2011) called Wen Tiejun's team "a group of people salvaging the dreams." Wen has worked at various governmental departments/think tanks regarding China's rural development and agricultural sector and is widely regarded as an official economist and the spokesman for peasants in China.

5 Members of these "farmers' cooperatives" are mainly peasants who are small-scale producers farming their plots under the Household Responsibility System. However, we use "farmers" rather than "peasants" here for two reasons: on the one hand, "farmers' cooperative" is an established and a more commonly used term compared to "peasants' cooperative"; on the other hand, these cooperatives are indeed producing for the market, rather than self-consumption.

6 The NRRM's attempts to connect producers with consumers started in 2006 when Guoren (Green Ground) Urban–Rural Cooperative was established to organize consumers to buy ecological produce from peasants across the country. Nevertheless, this cooperative could not be categorized as a CSA. CSA refers to a farm, not a cooperative. Our interview with an employee of Little Donkey Farm on April 4, 2012, indicated that shareholders of the CSA could also order food from remote peasants (e.g. from Lankao county, Henan province) through Green Ground.

7 The perceived threat of social stability by the government due to the gathering of people was also reflected in the operations of farmers' markets within cities. Our interviews with a buying club organizer on April 9, 2012, and two managers of the Beijing Farmers' Market, an ecological farmers' market which is a part of the NRRM, on April 3, 2012, December 6, 2012, and March 9, 2013, all demonstrated the challenge. See Chapter 7 for more discussion.

8 *Mu* is a Chinese measurement of farmland. 1 *mu* equals 0.164 acre.

9 Hazard-free is a type of certification along with "green" and organic certifications. On the differences of these three certifications, see Chapter 3.

10 The other four are economic, political, cultural, and social construction.

11 In interview a member of the NRRM working on a farm in Fujian, Fuzhou province, on June 2, 2012, told us:

> as long as we don't appeal to the higher authorities for rights, the government won't intervene in our work.... Our most important work right now is

to train a group of talented people who are familiar with the rural reconstruction values and can start this work in other parts of the country by
themselves even one day we [the NRRM] are dismissed.

12 On the role of NRRM in the development of AFNs in China, see Si and Scott
(2016).
13 Interviews with founders of CSA farms in Changzhou, Jiangsu province, on
April 14, 2012, Chengdu, Sichuan province, on April 26, 2012, Chongqing on
May 4, 2012, and Fuzhou, Fujian province, on June 2, 2012.
14 Interview with a leaseholder at Little Donkey Farm on April 4, 2012.
15 Interviews with a researcher in Beijing on March 29, 2012, and an intern at
Little Donkey Farm in Beijing on April 4, 2012.
16 A CSA farmer near Nanjing that we interviewed expressed her disappointment
about finding out about the support of government in the establishment and
operation of Little Donkey Farm. She said she was misled by the public image
portrayed by the media that Little Donkey Farm was pure civil society initiative, which led her to underestimate the difficulty in establishing and sustaining a CSA farm.
17 A buying club organizer we visited in Beijing on April 9, 2012, told us that
Little Donkey Farm was one of the only two farms they sourced food from
simply because it was a government-supported project and therefore was
trustworthy.
18 Our interview with the manager of the Big Buffalo Farm on April 15, 2012,
indicated that the local government invited Little Donkey Farm team to start
the project and built the infrastructure for the farm.
19 These terms refer to the rapidly growing agritourism that builds cultural and
recreational elements into agriculture and recreational plot rentals.
20 Interviews with a research in ecological agriculture in Beijing on March 29,
2012, and an intern working at Little Donkey Farm on April 4, 2012.

References

Ahlers, AL and Schubert, G 2009, " 'Building a new socialist countryside'—only a
political slogan?" *Journal of Current Chinese Affairs*, vol. 38, no. 4, pp. 35–62.
Central Committee of the CCP 2005, *Recommendations for Drafting the 11th Five
Year Plan.* Accessed at http://news.xinhuanet.com/politics/2005–10/18/
content_3640318.htm on November 22, 2013 (in Chinese).
Chen, B 2008, "The influence of changing peasant values on familial relations:
Liwei village, Anhui," *Chinese Sociology and Anthropology*, vol. 41, no. 1,
pp. 30–42.
Day, A 2008, "The end of the peasant? New Rural Reconstruction in China,"
Boundary 2, vol. 35, no. 2, pp. 49–73.
Day, A 2013a, *The Peasant in Postsocialist China: History, Politics, and Capit-
alism.* Cambridge University Press, Cambridge.
Day, A 2013b, "A century of rural self-governance reforms: Reimagining rural
Chinese society in the post-taxation era," *The Journal of Peasant Studies*, vol.
40, no. 6, pp. 929–954.
Day, A and Hale, MA 2007, "Guest editors' introduction," *Chinese Sociology and
Anthropology*, vol. 39, no. 4, pp. 3–9.
Démurger, S and Xu, H 2011, "Return migrants: The rise of new entrepreneurs in
rural China," *World Development*, vol. 39, no. 10, pp. 1847–1861.

Fan, CC 2008, *China on the Move: Migration, the State, and the Household*. Routledge, New York.

Goodman, D 2004, "Rural Europe redux? Reflections on alternative agro-food networks and paradigm change," *Sociologia Ruralis*, vol. 44, no. 1, pp. 3–16.

Guo, Y 2009, "Implications of Liang Shuming's rural reconstruction practice to the new socialist countryside construction," *Theory Monthly*, vol. 4, pp. 172–174 (in Chinese).

Hale, M 2013, *Reconstructing the Rural: Peasant Organizations in a Chinese Movement for Alternative Development*. Accessed at https://digital.lib.washington.edu/researchworks/handle/1773/23389 on December 4, 2013.

Hao, Z 2006, *The Role of Intellectuals in Rural Development in China: A Case Study of Pingzhou County in Shanxi Province*. Accessed at http://hdl.handle.net/123456789/12563 on October 21, 2013.

He, H 2007, "Experiments of new rural reconstruction in Lankao," *Chinese Sociology and Anthropology*, vol. 39, no. 4, pp. 50–79.

He, H, Cheng, X, and Zong, S 2014, "An empirical summary of and reflection on the contemporary New Rural Reconstruction Movement: A 10 years' experience in Kaifeng," *Open Times*, vol. 4. Accessed at www.opentimes.cn/bencandy.php?fid=376andaid=1827 on September 20, 2014 (in Chinese).

He, X 2007, "New rural construction and the Chinese path," *Chinese Sociology and Anthropology*, vol. 39, no. 4, pp. 26–38.

He, X 2008, "The regional variation of rural governance and the logics of peasant action," *Chinese Sociology and Anthropology*, vol. 41, no. 1, pp. 10–29.

Heilmann, S 2008a, "Policy experimentation in China's economic rise," *Studies in Comparative International Development*, vol. 43, no. 1, pp. 1–26.

Heilmann, S 2008b, "From local experiments to national policy: The origins of China's distinctive policy process," *The China Journal*, no. 59, pp. 1–30.

Ho, P and Edmonds, RL (eds.) 2007, *China's Embedded Activism: Opportunities and Constraints of a Social Movement*. Routledge, London.

Jacka, T 2013, "Chinese discourses on rurality, gender and development: A feminist critique," *The Journal of Peasant Studies*, vol. 40, no. 6, pp. 983–1007.

Liang, S 1936, "Our two big difficulties." In: *Complete Works of Liang Shuming*, Shandong People's Press, Jinan, pp. 573–585 (in Chinese).

Lin, H 2008, "The failure of amelioration?" *Southern People Weekly*. Accessed at http://news.sina.com.cn/c/2008–01–22/154214801573.shtml on December 13, 2013 (in Chinese).

Liu, S 2008, "The unity of rational construction and traditional heritage," *Jiangsu Social Sciences*, no. 1, pp. 141–146 (in Chinese).

Marsden, TK 2008, "Agri-food contestations in rural space: GM in its regulatory context," *Geoforum*, vol. 39, no. 1, pp. 191–203.

Marsden, TK, Banks, J, and Bristow, G 2000, "Food supply chain approaches: Exploring their role in rural development," *Sociologia Ruralis*, vol. 40, no. 4, pp. 424–438.

Marsden, TK, Banks, J, and Bristow, G 2002, "The social management of rural nature: Understanding agrarian-based rural development," *Environment and Planning A*, vol. 34, no. 5, pp. 809–825.

Marsden, TK and Sonnino, R 2008, "Rural development and the regional state: Denying multifunctional agriculture in the UK," *Journal of Rural Studies*, vol. 24, no. 4, pp. 422–431.

Nederveen Pieterse, J 2010, *Development Theory: Deconstruction/Reconstruction* (second edition). Sage, London.

Pan, J 2012, "Double movement: The rural reconstruction movement in China." PhD thesis, Lingnan University, Hong Kong, China. Accessed at http://commons.ln.edu.hk/cs_etd/17 on December 1, 2013 (in Chinese).

Pan, J and Du, J 2011a, "Alternative responses to 'the modern dream': The sources and contradictions of rural reconstruction in China," *Inter-Asia Cultural Studies*, vol. 12, no. 3, pp. 454–464.

Pan, J and Du, J 2011b, "The social economy of new rural reconstruction," *China Journal of Social Work*, vol. 4, no. 3, pp. 271–282.

Pei, X, Tandon, A, Alldrick, A, and Giorgi, L 2011, "The China melamine milk scandal and its implications for food safety regulation," *Food Policy*, vol. 36, no. 3 pp. 412–420.

Qiu, J 2007, "Rural reconstruction and the ruralization of knowledge: Qualms and hopes of the Zhaicheng experimental site," *Chinese Sociology and Anthropology*, vol. 39, no. 4, 80–96.

Renard, M and Guo, H 2013, "Social activity and collective action for agricultural innovation: A case study of new rural reconstruction in China," Accessed at http://hal.archives-ouvertes.fr/halshs-00802119 on December 4, 2013.

Renting, H, Marsden, TK, and Banks, J 2003, "Understanding alternative food networks: Exploring the role of short food supply chains in rural development," *Environment and Planning A*, vol. 35, no. 3, pp. 393–411.

Saich, T 2000, "Negotiating the state: The development of social organizations in China," *China Quarterly*, no. 161, pp. 123–141.

Schumilas, T 2014, "Alternative food networks with Chinese characteristics," PhD thesis, University of Waterloo.

Schneider, M 2015, "What, then, is a Chinese peasant? *Nongmin* discourses and agroindustrialization in contemporary China," *Agriculture and Human Values*, vol. 32, no. 2, pp. 331–346.

Scott, S, Si, Z, Shumilas, T, and Chen, A 2014, "Contradictions in state- and civil society-driven developments in China's ecological agriculture sector," *Food Policy*, vol. 45, pp. 158–166.

Shi, Y 2012, "De-labeling: romantic imaginations of big farms and small peasants," *Green Leaf*, vol. 11, pp. 37–43 (in Chinese).

Si, Z and Scott, S 2016, "The convergence of alternative food networks within 'rural development' initiatives: The case of the New Rural Reconstruction Movement in China," *Local Environment*, vol. 21, no. 9, pp. 1082–1099.

Southern Rural Daily 2005, "Interview with *sannong* scholar Wen Tiejun." Accessed at http://finance.sina.com.cn/review/20050715/19301803006.shtml on September 13, 2017 (in Chinese).

Tan, T 2007, "Paths and social foundations of rural graying," *Chinese Sociology and Anthropology*, vol. 39, no. 4, pp. 39–49.

Thøgersen, S 2009, "Revisiting a dramatic triangle: The state, villagers, and social activists in Chinese rural reconstruction projects," *Journal of Current Chinese Affairs*, vol. 38, no. 4, pp. 9–33.

Tong, D 2006, "Congratulations to Professor Wen Tiejun for not being able to sell his rice," *China Youth Daily*. Accessed at http://edu.people.com.cn/GB/1055/4019861.html on December 13, 2013 (in Chinese).

Tovey, H 2009, "'Local food' as a contested concept: Networks, knowledges and power in food-based strategies for rural development," *International Journal of Sociology of Agriculture and Food*, vol. 16, no. 2, pp. 21–35.

Wang, H 1998, "Contemporary Chinese thought and the question of modernity," *Social Text*, no. 55, pp. 9–44.

Wang, L 2006, "Professor He Huili: Why do I have to sell rice?," *China Youth Daily*. Accessed at http://news.xinhuanet.com/school/2006–01/17/content_4060 662.htm on December 14, 2013 (in Chinese).

Wang, RY, Si, Z, Ng, C, and Scott, S 2015, "The transformation of trust in China's alternative food networks: Disruption, reconstruction and development," *Ecology and Society*, vol. 20, no. 2, article 19.

Wang, X 2009, "Seniors' organizations in China's new rural reconstruction: Experiments in Hubei and Henan," *Inter-Asia Cultural Studies*, vol. 10, no. 1, pp. 138–153.

Watts, DCH, Ilbery, B, and Maye, D 2005, "Making reconnections in agro-food geography: Alternative systems of food provision," *Progress in Human Geography*, vol. 29, no. 1, pp. 22–40.

Wen, T 2007, "Deconstructing modernization," *Chinese Sociology and Anthropology*, vol. 39, no. 4, pp. 10–25.

Wen, T 2009, *The "Sannong" Issue and Institutional Changes*. China Economic Publishing House, Beijing (in Chinese).

Wen, T and Lau, K 2008, "Four stories in one: Environmental protection and rural reconstruction in China," *East Asia Cultures Critique*, vol. 16, no. 3, pp. 491–505.

Wen, T, Lau, K, Cheng, C, He, H, and Qiu, J 2012, "Ecological civilization, indigenous culture, and rural reconstruction in China," *Monthly Review*, vol. 63, no. 9, pp. 29–44.

Wen, T and Sun, Y 2012, "Two transformations at the turn of the century and a new interpretation of the *sannong* issue," *Inquiry into Economic Issues*, vol. 9, pp. 10–14 (in Chinese).

Weng, S 2008, "The shutdown of rural reconstruction college: The ending of 'new Dingxian experiment?'" Accessed at www.eeo.com.cn/eeo/jjgcb/2008/01/07/ 90690.shtml on December 13, 2013 (in Chinese).

Wu, Y, He, X, Luo, X, Dong, L, and Wu, L 2008, "The path and subject of rural governance studies: A response to Ying Xing's critique," *Chinese Sociology and Anthropology*, vol. 41, no. 1, pp. 57–73.

Wu, Y and Li, D 2008, "Twenty years of rural political studies: The rise and fall of a public academic movement," *Chinese Sociology and Anthropology*, vol. 41, no. 1, pp. 74–99.

Xu, J 2000, "The upheaval between radicalness and conservativeness." In: Li, S (ed.), *The Standpoint of Intellectuals: The Upheaval between Radicalness and Conservativeness*. Time Literature and Art Press, Changchun (in Chinese).

Xu, N 2011, "From James Yen to Little Donkey Farm," *China Fortune*. Accessed at http://gcontent.oeeee.com/chinaFortune/b/d4/bd4c9ab730f55132/Article.html on December 13, 2013 (in Chinese).

Yan, H and Chen, Y 2013, "Debating the rural cooperative movement in China, the past and the present," *The Journal of Peasant Studies*, vol. 40, no. 6, pp. 955–981.

Yeh, E, O'Brien, K and Ye, J 2013, "Rural politics in contemporary China," *The Journal of Peasant Studies*, vol. 40, no. 6, pp. 915–928.

Ying, X 2008, "Critique of a new trend in villager self-government studies," *Chinese Sociology and Anthropology*, vol. 41, no. 1, pp. 43–56.

Yu, Y 2006, "Radical ideas and conservative ideas in modern China." In: Xu, J (ed.), *Historical Ideologies in China in the 20th Century* (second edition). Orient Press, Shanghai (in Chinese).

Yu, Y 2011, "The return of white collar workers to rural reconstruction: Putting an end to the romantic imagination of rural life." Accessed at http://zylz.net/zixun/qingchun/3521_2.html on December 13, 2013 (in Chinese).

10 Conclusion

Zhenzhong Si, Steffanie Scott,
Theresa Schumilas, and Aijuan Chen

China has undergone a tumultuous 70 years of agrofood system reforms and food security challenges, most notably the great famine of 1958–1961. Overcoming this devastating crisis and putting itself on a more secure footing in terms of food production has been a huge achievement. Yet, over the last three decades, the damaging ecological and social consequences of boosting production have become ever more evident. China's agriculture sector faces huge environmental challenges—from overuse of chemical fertilizers, to water pollution and soil erosion—and social and health challenges, from widespread food safety concerns (from agrochemical residues in food to adulteration of processed foods, leading to a crisis of trust) to the decline of the agricultural labour force and hollowing-out of rural communities (Lu et al. 2015; Wang et al. 2015; Zolin et al. 2017; Pretty and Bharucha 2015; Luan et al. 2013; Fang and Meng 2013; Han 1989; Campbell et al. 2017).

Diversified, ecological agriculture systems offer tremendous opportunities to address these concerns by "replacing chemical inputs, optimizing biodiversity and stimulating interactions between different species, as part of holistic strategies to build long-term fertility, healthy agro-ecosystems and secure livelihoods" (Friesen 2016: 3). To this end, China has seen a range of government, farmer-led, grassroots, and consumer-led initiatives over the past decade or more. Government policies have been enacted for organic and "green" food quality certifications. Consumers have sought safer and more ecologically grown foods. Some farms have adapted to meet these demands, through achieving certifications or selling uncertified products directly to consumers. And grassroots organizing has taken shape to establish organic farmers' markets and buying clubs. All of this has had some impact, but the problems continue to loom large. China's organic agriculture sector in particular, and its food system in general, face critical challenges that hinder their future sustainability:

1   Small organic farmers are largely excluded from organic certification.
2   The value of traditional agroecological practices are overlooked by state planners and society overall.

3   Traditional farmers are unable to earn price premiums from foods that they grow without agrochemicals.
4   Organic sector support organizations are lacking.
5   Organic agriculture is perceived by state officials and others to have low productivity, and thus officials are reticent to promote it widely.
6   Organic and ecologically produced foods are unaffordable for most consumers.
7   Research on organic production techniques in China is quite limited.

These challenges highlight the need for a holistic food systems perspective in China to expand the ecological agriculture sector in a more ecologically and economically sustainable direction.

Our research has underscored the diversity and complexity of development paths of organic agriculture in China. We sketched the broad socio-political landscape in China that is shaping the food policy regime under which the ecological agriculture sector and AFNs are emerging and evolving. Our analysis revealed the relevance of various political economy factors (e.g. state roles, civil society constraints, the land tenure system, and the emergence of the middle and upper-middle classes) in explaining the development path of certified organic agriculture, and ecological agriculture more widely, in recent decades in China.

We characterized China's complicated food policies as a system centring on national food security, or, more explicitly, sufficient grain productivity. This policy regime dominated by food security is reflected by various policies revolving around farmland preservation, food reserve and circulation governance, and agricultural policy support. It is also increasingly complemented by concerns of food safety and environmental sustainability of food production. This characterization of China's food policy regime is important for understanding the state's priority of "agricultural modernization." It also provides the context for understanding many of the state impediments and initiatives in ecological agriculture development in China. Part of this analysis also included working out the various types of involvement of small-scale farmers in this sector, including through AFNs. The findings about the different mechanisms for involvement of small-scale farmers in the organic agriculture sector, and the implications of these for farmers, point to the need for policy to better support small-scale farmers in ecological agriculture.

Our overall objective was to understand China's ecological agriculture sector in terms of both top-down and bottom-up initiatives. In particular, we sought to examine organic agriculture and green food farms that received state support; private sector–initiated organic farms; farmers' professional cooperatives conducting ecological agriculture; and civil society–led alternative food initiatives and their association with the New Rural Reconstruction Movement. Some of the foci of our analysis of top-down EA initiatives examined the ownership structures (who is doing organic farming in China), the involvement of small-scale and "traditional" farmers, the implications of farmers' cooperatives for sustainable rural development, and the challenges and distribution of economic benefits within farmers' cooperatives. For the bottom-up initiatives, we elaborated a typology of AFNs, their key principles, inherent values, and internal contradictions. We highlighted the social and political conditions that shape AFNs, similarities and differences between AFNs in China and the West, AFNs' roles in achieving broader rural development goals, and organizing strategies for food system transformation.

State roles and AFN development

The book focuses on the dual top-down and bottom-up approaches to building ecological agriculture in China. The differences between the top-down and bottom-up developments complicate the interactions between the state and civil society actors in shaping the developmental path of ecological agriculture. While many CSA farms are interested in receiving more support from the state, they operate with a wider set of values, which diverge from the state's agenda in promoting modern agriculture. The combination of top-down and bottom-up approaches operating in China also reduces the transformative potential of organic agriculture development since the state's support for "mainstream" ecological agriculture farms and interventions in AFNs diminishes the latter's space of action to make system changes.

The state's role in top-down initiatives for ecological agriculture development distinguishes China's organic agriculture sector from other countries. Organic agriculture in the West has been mainly initiated by individual farmers, and sometimes non-governmental organizations, in response to the challenges caused by conventional farming systems. In developing countries and the BRICs, organic agriculture has been mainly promoted by exporters, with a significant role played by civil society organizations in some countries (e.g. India and Brazil). Outside Europe, states have generally played limited roles in the organic agriculture sector. In contrast, the Chinese government has played an important and strong role in organic agriculture development. One of these involvements is reflected in the establishment of a set of food production standards (hazard-free, green, and organic). Central and local government agencies in China have also played pivotal roles in developing the organic agriculture sector by working closely with the private sector, implementing favourable policies, providing various subsidies and extension services, and helping establish market linkages. More specifically, the Chinese government has been playing an important role in establishing a supportive environment for farmers' cooperatives, mainly through (1) implementing the Cooperative Law and developing a series of favourable policies, (2) intervening directly in the establishment and operation of cooperatives, and (3) providing various forms of financial support (e.g. subsidies, tax exemption and preferential loans) and non-financial support (e.g. technical assistance, marketing, and public recognition). However, this support has been disproportionately channelled toward larger farms.

Our research also revealed the critical role of the state in shaping AFN developments in terms of setting restrictions and challenging legitimacy. The state is also a key player in shaping the contested nature of AFNs, as exemplified in the case of the Beijing Farmers' Market. Just like the state's role in top-down approaches to ecological agriculture development, the context of having the state as a critical player affecting the practices of AFNs in China also differs remarkably from AFN development in the West. Recognizing the potential of AFNs in generating new economic opportunities, the state also fosters the development of AFN in certain circumstances. This is reflected, for example, in the establishment of several CSA farms with the support of local governments.

The Chinese state's complicated role in AFN development is also reflected in its unaltered policy orientation that falls into the "agroindustrial" rural development paradigm. The deep-rooted pro-industrial sentiment is expressed in guiding governmental policy documents. The Chinese state strives to achieve sustainability within an industrial model of agriculture development. While proclaiming carbon emission reductions within "modern agriculture" and environmental protection of farmland, the state continues to interpret "agricultural modernization" as high-efficiency, high-tech, large-scale, mechanized, standardized, professionalized, and productivity-oriented agriculture. Nevertheless, the state's promotion of new developmental discourses such as "ecological civilization," "resource efficient and environmentally friendly agriculture," and "low carbon economy" also bestows a certain degree of legitimacy and significance on AFN development in China.

Indeed, this state's contradictory position with respect to industrial and ecological agriculture contributes to its complicated and contradictory role in AFN development. As environmental consequences of the agriculture sector are increasingly recognized, the Chinese government has issued various policies to promote green and sustainable agriculture in the past few years, such as zero growth of chemical fertilizer and pesticide use by 2020 (see Jin and Zhou 2018). Yet, these recent emphases on agricultural sustainability have not completely altered its long-lasting policy orientation that supports agriculture industrialization. The productivist agricultural and rural development policies, as Wen Tiejun argues (see Wen and Sun 2012), help little in solving the *sannong wenti* (three rural problems). Rather, such policies are increasingly contributing to economic, social, and environmental crises in the countryside. Therefore, as Wen Tiejun's new interpretation of *sannong wenti* suggests, there needs to be a shift of policy focus from increasing productivity and boosting farmers' incomes to rural sustainability, ecological safety, and protection of farmers' rights. This resonates with the emergence of the "rural development" paradigm that has been observed in Europe, although there is still a long way to go before this could be translated into policies and practice in China.

Characterizing Chinese AFNs

Drawing on the extensive discussions of AFNs and rural development in the West, our analysis examined AFNs from different angles. One finding of our study is that CSAs, farmers' markets and buying clubs—initiated by a diversely motivated group of primarily young, educated urbanites—are rapidly expanding and creating a space for themselves in the world's largest food economy. Shaped by strong imaginaries of traditional Chinese agriculture, these AFNs are economically diverse and reveal reconnections between people and land.

Driven by the growing awareness of the ecological and social implications of consuming "good" food, Chinese consumers in the years to come will have an increasing demand for food channelled through AFNs. Although there is a value distinction between AFN initiators and their customers, we expect that this distinction will be gradually minimized with these initiators' efforts in public education. Indeed, this will be a long and rough process but we have seen actions being taken. Food education is increasingly being recognized by the public. Food courses in the field are happening on Shared Harvest Farm in Beijing. Although AFNs will be marginal in China's agrifood system for a long time, they are engaging more and more people.

Our research also probed into the discussion of alternativeness in AFNs. The economic, social, ecological, and political dimensions of alternativeness have been criticized in analyses of AFNs in the West. Debates over the extent to which AFNs are alternative shook the foundation of AFNs' counter-conventional stances. These critiques cast doubt on the "transformative potential" of AFNs (see Levkoe 2011) in terms of generating a meaningful alternative to the conventional food system. At the same time, they highlighted the fact that AFNs were not necessarily alternative in the dimensions of this term that we might consider. We argue that it is too simple to negate the status of certain AFNs because they lack certain dimensions of alternativeness. Instead, we posit an examination of these initiatives according to an unpacked and more specific set of elements of alternativeness. These elements, which represent various alternative features of food and of embedded relations, offer a more feasible analytical framework for characterizing AFNs, especially nascent ones that have not developed a full spectrum of alternativeness. In this way, our research demonstrates that the landscape of alternativeness varies among different AFNs. Rather than simply categorizing an initiative into either alternative or conventional, we offered examples of the various ways in which AFNs can be alternative.

We argue that AFNs in China are evolving rapidly due to several reasons. First, the nascency of these AFNs makes it all the more difficult to anticipate their future trajectories. Most of these initiatives only emerged after 2008. Their value system and modes of operating are fluid and still being defined. Second, the rapidly changing social trends will further complicate the landscape of AFNs. AFNs have flourished in some parts of China because of the food safety crisis of 2008 and its aftermath. Whether and for how long this food safety anxiety will continue to fuel the expansion of AFNs is unclear. Meanwhile, as the NRRM is functioning as a driver for the development of AFNs, the expansion of AFNs seems to be increasingly values-oriented. Thus, addressing food safety concerns might become less important in the future. Third, the changing power dynamics among various groups, including food activists, grassroots organizations, consumers, ecological producers, and the state, will also shape their developmental paths. Food activists are promoting ethical consumerism and the ecological and social values of AFNs. The state is promoting modern industrialized agriculture while also giving somewhat more emphasis to ecological sustainability. Consumers are looking for healthy and safe food while receiving information from both food activists and the state. Thus, how and to what extent AFNs will remain alternative in China is highly dependent upon the dynamics among these key actors. Therefore, some findings in this book, such as the landscape of alternativeness of various AFNs, may fluctuate in the future. Despite this, by depicting their emergence and characterizing these initiatives, our study provides "stepping stones" for further studies of the evolution of AFNs in China.

Political dimensions of AFN development

In this book, we also explore AFNs' complicated reciprocal relationship with the rural development initiative—the NRRM. Although AFNs were introduced to China by food activists from the West, the Chinese versions are significantly different from their counterparts owing to the strong consumer-driven feature and the interventions of the state in its development. The evolving manifestations of alternativeness, the inconsistency in ethical values between their initiators and customers, as well as the contestations among various players, all differentiate AFNs in China from their counterparts in the West. Thus, AFNs in China have been significantly shaped by the specific socio-political contexts.

Regarding the relationship between the NRRM and AFNs, our examination of the NRRM in China, especially challenges confronting it and its strategies to address these challenges, offered an empirical example to illustrate another layer of the role of AFNs in rural development. AFNs thus can be not only "forerunners," "engines," and "building blocks" (Renting et al. 2003; Goodman 2004; Marsden et al. 2000) of the emerging new rural development paradigm (see van der Ploeg et al. 2000; Pugliese 2001;

Marsden et al. 2002; Goodman 2004) but also a powerful tool to address challenges of achieving sustainable rural development.

The complexity of the NRRM's adaptation to state power and societal conditions illustrated the importance of flexibility in a grassroots rural development initiative. In a political environment in which collective actions are highly sensitive, it is especially a challenge to constitute new forms of social organizations. Marsden (2008) has pointed out that the "rural development" model can only become influential with a "new policy support structure." Marsden et al. (2002: 817) also argue that "new agrarian-based rural development" will not become mainstream unless it is "accompanied by a redefined and strategically organized rural development policy framework." In this sense, although the NRRM is an initiative that represents an alternative rural development model, the paradigm has not been formalized in China. The agrarian-based rural development model, which differs from the "agroindustrial model," is still at its early stage of formation. The NRRM will have to strive for policy impacts before it can facilitate the rural development paradigm shift in China.

It is also worth highlighting that the concept of "local food" was rarely mentioned during our interviews. This contrasts with the practices and research around AFNs in the West, where "the local" is either reified and/or critiqued. While almost all of the AFNs we examined were procuring and exchanging local food, they were not pursuing a local ideology that reifies scale, nor were they local out of necessity. There is a pragmatic, rather than utopian, understanding of place in these networks. In contrast to "defensive localism," and perhaps influenced by China's experience of "opening to the world," there is a strong orientation to search for knowledge, experience, and information from beyond China and remix these with "Chinese characteristics." The ways in which China's AFNs practise a politics that is rooted in place but also looks outward supranationally warrants further research.

Another dimension of the political implications of AFNs in China relates to their transformative potential. Similar to their counterparts in the West, Chinese AFNs can be blind to privilege and perpetuate some of the very injustices they seek to transform. Yet, we argue that, by using inclusive and reflexive processes, participants are building diverse networks that hold transformative potential. In contrast to AFNs in the West, however, in the context of pervasive uncertainty of an authoritarian state, Chinese AFNs have adopted a subtle everyday resistance style (Scott 1985). This is reflected in the critiques of government support ignoring small peasant farmers and related food initiatives, frustration with the state's food safety governance, and many other disguised examples of everyday resistance. These AFNs are actively, though not always with full awareness perhaps, positioning themselves as a path or a "portal" to building connections to broader emancipatory spaces of global social justice movements.

As China's AFNs join others from the West in "moving beyond the market," they reveal new ways to think about resistance. They challenge the assumption in AFN theorizing of an independent civil society sphere where non-state actors can gather, discuss, and challenge policy. The examples described here call for an extension to the ways in which AFN scholarship understands citizenship and suggest that actions for the common good can take place at all scales, from the personal to the global, as well as through diverse styles—from the overt to the everyday. China's AFNs reveal that hegemony is never total. Even in the absence of organized civil society and in a context of pervasive uncertainty, resistance finds its space. The everyday resistance repertoires detailed here suggest that space, between domination and overt defiance, is large, and worthy of further exploration in different contexts. It could be fruitful to bring theories of everyday resistance into theorizing AFNs in the West.

Integration of small-scale farmers

By investigating the participation of small-scale farmers in different models of top-down initiatives, our research shows that they have played an important role in linking smallholders to value-added markets and have contributed to boosting farmers' incomes. We recognize, however, that China's unique land tenure system—featuring the separation of land ownership into contractual and operation land rights (Wang and Zhang 2017)—impacts the extent and type of involvement of small-scale farmers in the organic agriculture sector. Unlike the contract farming and land leasing models in which enterprises play a dominant role and make most (if not all) decisions, farm members in the independent cooperative model have played an active role in decision-making and enjoyed more autonomy. Farmers in the independent cooperative model had a holistic understanding of organic agriculture and demonstrated a stronger commitment to sustainable development in their daily operations. The model provides small-scale farmers with more opportunities to improve their livelihoods and ensures a long-term viability of local communities by involving small-scale farmers in decision-making. However, the case study also found that economic gains were not shared equally among members in the cooperative. "Common" members only benefited through selling their products to the cooperative, whereas "core" members benefited through both selling their products and investing capital in the cooperative.

In contrast to the experience of farmers' cooperatives, China's AFNs privilege connecting to land and urban entrepreneurs who operate farms, over connecting to the peasants who grow the food and labour on these farms. In fact, it is not only the consumers in these networks who display a distrust of peasant farmers; AFN organizers and CSA entrepreneurs at times also seem to reinforce the marginalization of peasants. For some CSA operators in these networks, peasant farmers are simply labour, and there is no attempt to integrate them into the decision-making on the farms. When asked about the involvement of peasants in the farms, these organizers replied that the peasants had lost traditional farming skills and that they would have very little to share in planning the work on the farm. This is an interesting perspective considering that peasants come from families with centuries of experience working the land, while the urban people starting these CSAs are often new to farming. Indeed, those CSA operators who come from urban rather than peasant backgrounds seemed blind to this "othering" and sometimes appeared more concerned about the availability of "cheap labour" than about supporting recent state policies that address rural marginalization. Suffice to say that China's AFNs can sometimes reinforce social injustices based on entrenched inherited inequities. Certainly, there are efforts to address injustices in these networks through charitable acts. Farmers' markets use money raised from food sales to purchase food for peasants living in poor districts and subsidize peasant farmers to attend training events and workshops. However, these localized approaches or "band aids" do not fundamentally challenge structural conditions or cultural discourses, such as *suzhi*, that perpetuate marginalization.

Limitations of our study

As with any study, there are some limitations with what we were able to accomplish. One limitation with using interviews as a key method for data collection relates to interviewees' intentional avoidance of politically sensitive topics. One of our goals was to understand the roles of the state in relation to AFN development. However, criticizing the state has always been risky in China. Thus, our informants might have refrained from expressing their viewpoints. In one of our interviews with a buying club organizer, the interviewee was stopped by his colleagues (a fellow buying club organizer) for criticizing the government, and the conversation was switched to a different topic. In order to understand how the government has put pressure on the operation of farmers' markets, we explicitly asked our informants about this and sometimes emailed them afterwards, yet the responses were still obscure. Another limitation of interviewing is the tendency for interviewees to report only the "good" (success stories) but not the "bad" (challenges). This can be a problem especially if the informants treated us (with the presence of Steffanie Scott and Theresa Schumilas) as foreign quasi-journalists who could potentially promote their initiatives.

In addition, the snowball/convenience sampling method might bring us only one side of the story and lead to a biased selection of interviewees. This is because interviewees may only refer us to people with similar opinions, and certain groups of people might be left out. In this sense, the study will tend to reflect perspectives of a certain group of people with similar interests. This is especially the case for the earlier interviews conducted with state-supported top-down initiatives; interviewees and farm visits were sometimes identified by local government officials and/or certification agencies. Participants could be more easily recruited and approached with recommendations and assistance from these officials or staff members. In some cases, these officials or staff made a call to help us arrange a visit. In a few other cases, interviews and site visits were conducted in the presence of local officials. Although participants did not feel compelled to be interviewed if government officials had recommended them, the presence of local officials during the interviews and site visits could influence interviewees' responses (e.g. saying positive things about the government and being silent about the negatives), especially when the officials were in charge of the sector of ecological agriculture in the region. Therefore, the positive roles of government agencies might be overstated, while their negative impacts and shortcomings might be overlooked. Despite this, it was still helpful to have a local guide to lead us to appropriate participants who in some cases might not have been possible to access on our own. This sampling approach also affected the farms that we were directed to, some of which were "model farms" that had gained government support and were relatively profitable.

The timing of the fieldwork is also important as it affects whom we could interview. Some of our interviews and site visits were made in slack farming seasons when many part-time farmers had migrated to urban areas for work. Thus, farmers that Aijuan Chen interviewed were mainly elderly people. This created bias because younger farmers sometimes have different understandings of ecological and organic agriculture. As one of our interviewees noted, young farmers are more interested in ecological farming, while older farmers are more conservative and reluctant given the risk associated with and labour required for ecological agriculture.

To overcome some of the limitations mentioned above, we used multiple sources of data to triangulate observations and interpretations. We also drew on contacts from a variety of sources (from government, academics, certification agency sources, CSA farmer networks, and businesses) to avoid over-relying on responses from just one group of people with similar ideas. Besides snowball sampling, some of us followed blogs of people who disagreed with (and publicly criticized) our interviewees. We also integrated perspectives and issues from online forums and microblogs that were not captured in our interviews. Finally, news reports were also taken into consideration to make up for the potential one-sided viewpoints represented by some interviewees.

Future research

Given the exploratory nature of our study, there are ample avenues for follow-up research. Future research on farmers' cooperatives in China's ecological agriculture sector could examine both successful and less successful cases to shed more light on the obstacles that cooperatives have encountered as well as the elements of success. In our study we found that cooperative benefits were not equally distributed among members because of differences in assets and resources. It would be insightful for future studies to explore whether there is a connection between these different "classes" of membership and the extent of decision-making in and economic benefits from farmer professional cooperatives. Such research could challenge assumptions about how equitable these cooperatives are in practice.

Another avenue for follow-up research relates to the characteristics of AFNs as a function of specific socio-economic and political conditions. Chapter 7 explored various contestations within a farmers' market in Beijing in terms of the power structure, vendor relationships, consumer motivations and the discourse of "local." Nevertheless, these contestations are place-based and context-specific. Other AFNs, especially CSAs and recreational garden plot rentals in China, are also situated within local conditions. Understanding this situatedness is key to understanding the diversity and evolution of AFNs in China. This requires more empirical case studies that examine the values and operational challenges confronting these initiatives.

For our research on AFNs, although our work was based mainly on interviews with initiators of these ventures, consumer perspectives are vital. We did hear about consumer opinions, particularly their desire for safe and healthy food, in our interviews with initiators of AFNs and though secondary sources. However, further research on consumer perspectives will beneficial to sketching out the whole picture of AFNs in China. This makes Chen Weiping's past (2013a, 2013b) and current investigations of the features of Chinese CSA customers, and their engagement through social media, important.

In addition, understanding the reciprocal relationship between AFNs and the NRRM, or rural development initiatives in general, requires questioning the impacts of rural development initiatives on the development of AFNs. Although a separate paper (Si and Scott 2016) addresses this issue by examining the NRRM's efforts in the convergence of dispersed AFNs, further studies are needed to monitor the impacts on food policies and food systems in China. That is, how will AFNs' community building and capacity enhancement through NRRM be translated into structural changes?

Paths forward

After more than three decades of rapid development, China, as the world's second largest economy, is at a critical stage of transitioning toward a sustainable development path. Having recognized the critical role of agriculture in this transition, the Chinese government has indicated its support for sustainable agriculture in various governmental documents and policies. Yet, as we have identified previously in this book, critical gaps still exist in these policies that isolate agriculture from other broad food system challenges, exclude smallholders from joining the greening of agriculture, sideline social justice concerns, and diminish the political space for social organizations. Thinking beyond ecological agriculture to food system sustainability will be key for China to address a myriad of ecological and socio-economic challenges. AFNs offer important ways forward in this vein.

It is also increasingly clear that ecologically produced food is about more than just agriculture and an issue that pertains to the food system. The greening of the agriculture sector has clear connections to development agendas such as poverty reduction and rural development that are well recognized by the Chinese government. Other connections, such as gender equity or climate change, are also highly pertinent to ecological agriculture development but may not be widely acknowledged within China. Moreover, there is little discussion in China about the United Nations Sustainable Development Goals (SDGs). The SDGs lay out a new and universal developmental agenda that UN member states are expected to use to frame their policies between 2015 and 2030. Agriculture is, in fact, one common thread holding the SDGs together (Farming First 2015), and food production following agroecological principles is closely connected with most of the SDGs (Farrelly 2016; FAO 2016). A quality food supply, and one that avoids or reduces the use of synthetic fertilizer, helps to ensure a cleaner water supply and biodiversity, which in turn enhances SDG #2 Sustainable cities and communities, SDG #14 Life below water, and SDG #15 Life on land, to mention just a few. Fully recognizing these connections will ensure a more holistic and sound mechanism to facilitate food policymaking.

Thinking beyond China, we believe China's experience offers some lessons for other countries seeking to develop their ecological agriculture sectors. State support to cover certification costs, for example, would be a welcome help for farmers elsewhere. Beyond state support, there may be lessons from CSA farms' use of social media to engage their customers and share info about their products and vision and values. Consumers, for their part, have been active in using social media platforms to share their concerns and insights about food safety and about trustworthy food sources. Meanwhile, China also needs to learn from other countries' experience in community building, skill training, and social engagement. It is therefore imperative to enhance partnerships at various political and spatial levels for knowledge mobilization and information sharing.

Looking to the future, it is difficult to say affirmatively whether China's ecological agriculture initiatives will be mainstreamed or co-opted, against the backdrop of the negotiation between the state, the private sector and civil society actors. However, one thing is for sure: your daily food choices matter and the future—of agricultural sustainability and our global and local food systems—are in our hands, and on our chopsticks and forks.

References

Campbell, BM, Beare, DJ, Bennett, EM, Hall-Spencer, JM, Ingram, JSI, Jaramillo, F, Ortiz, R, Ramankutty, N, Sayer, JA, and Shindell, D 2017, "Agriculture production as a major driver of the Earth system exceeding planetary boundaries," *Ecology and Society*, vol. 22, no. 4, pp. 8.

Chen, W 2013a, "Perceived value in community supported agriculture (CSA): A preliminary conceptualization, measurement, and nomological validity," *British Food Journal*, vol. 115, no. 10, pp. 1428–1453.

Chen, W 2013b, "Perceived value of a community supported agriculture (CSA) working share: The construct and its dimensions," *Appetite*, vol. 62, pp. 37–49.

Fang, L and Meng, J 2013, "Application of chemical fertilizer on grain yield in China analysis of contribution rate: Based on principal component regression C-D production function model and its empirical study," *Chinese Agricultural Science Bulletin*, vol. 29, no. 17, pp. 156–160 (in Chinese).

FAO. 2016, "Soils and pulses: Symbiosis for life." Accessed at www.fao.org/docu ments/card/en/c/56244a4c-d35a-48f8-b465-89f46f343312 on May 9, 2018.

Farming First. 2015, "The story of agriculture and the sustainable development goals." Accessed at https://farmingfirst.org/sdg-toolkit#section_2 on May 9, 2018.

Farrelly, M 2016, "Agroecology contributes to the sustainable development goals," Ileia. Accessed at www.ileia.org/2016/09/22/agroecology-contributes-sustainable-development-goals on May 9, 2018.

Friesen, S 2016, *From Uniformity to Diversity: A Paradigm Shift from Industrial Agriculture to Diversified Agroecological Systems*. Brussels: International Panel of Experts on Food Systems (IPES-Food).

Goodman, D 2004, "Rural Europe redux? Reflections on alternative agro-food networks and paradigm change," *Sociologia Ruralis*, vol. 44, pp. 3–16.

Han, C 1989, "Recent changes in the rural environment in China," *Journal of Applied Ecology*, vol. 26, pp. 803–812.

Jin, S and Zhou, F 2018, "Zero growth of chemical fertilizer and pesticide use: China's objectives, progress and challenges," *Journal of Resources and Ecology*, vol. 9, no. 1, pp. 50–58.

Levkoe, CZ 2011, "Towards a transformative food politics," *Local Environment*, vol. 16, no. 7, pp. 687–705.

Lu, Y, Jenkins, A, Ferrier, RC, Bailey, M, Gordon, IJ, Song, S, and Zhang, Z 2015, "Addressing China's grand challenge of achieving food security while ensuring environmental sustainability," *Science Advances*, February, pp. 1–5.

Luan, J, Qiu, H, Jing, Y, Liao, S, and Han, W 2013, "Decomposition of factors contributed to the increase of China's chemical fertilizer use and projections for future fertilizer use in China," *Journal of Natural Resource*, vol. 28, no. 11, pp. 1869–1878 (in Chinese).

Marsden, TK 2008, "Agri-food contestations in rural space: GM in its regulatory context." *Geoforum*, vol. 39, no. 1, pp. 191–203.

Marsden, TK, Banks, J, and Bristow, G 2000, "Food supply chain approaches: Exploring their role in rural development," *Sociologia Ruralis*, vol. 40, no. 4, pp. 424–438.

Marsden, TK, Banks, J, and Bristow, G 2002, "The social management of rural nature: Understanding agrarian-based rural development," *Environment and Planning A*, vol. 34, no. 5, pp. 809–825.

Pretty, J and Bharucha, ZP 2015, "Integrated pest management for sustainable intensification of agriculture in Asia and Africa," *Insects*, vol. 6, pp. 152–182.

Pugliese, P 2001, "Organic farming and sustainable rural development: A multifaceted and promising convergence," *Sociologia Ruralis*, vol. 41, no. 1, pp. 112–130.

Renting, H, Marsden, TK, and Banks, J 2003, "Understanding alternative food networks: Exploring the role of short food supply chains in rural development," *Environment and Planning A*, vol. 35, pp. 393–411.

Scott, J 1985, *Weapons of the Weak: Everyday Forms of Resistance*, Yale University Press, New Haven, CT.

Si, Z and Scott, S 2016, "The convergence of alternative food networks within 'rural development' initiatives: The case of the New Rural Reconstruction Movement in China," *Local Environment*, vol. 21, no. 9, pp. 1082–1099.

van der Ploeg, JD, Renting, H, Brunori, G, Knickel, K, Mannion, J, Marsden, T, Roest, K de, Sevilla-Guzmán, E, and Ventura, F 2000, "Rural development: From practices and policies towards theory," *Sociologia Ruralis*, vol. 40, no. 4, pp. 391–408.

Wang, Q and Zhang, X 2017, "Three rights separation: China's proposed rural land rights reform and four types of local trials," *Land Use Policy*, vol. 63, pp. 111–121.

Wang, X, Cai, D, Grant, C, Willem, B, and Hoogmoed, O 2015, "Factors controlling regional grain yield in China over the last 20 years," *Agronomy for Sustainable Development*, vol. 35, no. 3, pp. 1127–1138.

Wen, T and Sun, Y 2012, "Two transformations at the turn of the century and a new interpretation of the Sannong issue," *Inquiry into Economic Issues*, no. 9, pp. 10–14 (in Chinese).

Zolin, MB, Cassion, M, and Mannino, I 2017, *Food Security, Food Safety and Pesticides: China and the EU Compared*. Working paper, Ca' Foscari University of Venice.

Epilogue
Anecdotes of fieldwork – self-reflections from our diverse positionalities

Steffanie Scott, Zhenzhong Si, and Theresa Schumilas

It has been a privilege and a delight to work with such a capable and committed group of co-authors (all former PhD students of mine). The learning and dialogues between Theresa and me, as Canadians familiar with the organic and wider food movements and sector development, and Aijuan and Zhenzhong, who grew up in China, were profound. I remember, on an overnight train ride from Beijing to Dalian (Shenyang province), Theresa and I explained to Zhenzhong the concepts of a "hippie" and the "back-to-the-land" values of the counter-cultural generation of the 1960s and 70s that spawned a generation of pioneer organic farmers in North America. This cross-fertilization of ideas and experiences between North America and a Chinese context of historical experience and institutional and sociocultural practices, has enriched us all. We particularly enjoyed presenting on our Canadian organic food system experiences, strengths, and limitations to vendors and volunteers at the Beijing Farmers' Market and at the Organic Food Development Centre.

It felt like a dream come true to travel around China from one ecological farm to another, taking in the diversity of the landscape, crops, farming practices, marketing strategies, and motivations for farming in each case. As Zhenzhong notes below, the contrast in orientations and styles of farm operations was striking. At one extreme, I took a bus for four or five hours, then was picked up and driven an hour through bamboo forests to the end of a road, beside a lake, where we were escorted into an ornately decorated private boat and ferried across, where we were met by someone in a golf cart and taken up a small hill to an estate that felt like a movie set. Everyone was dressed in Qing dynasty robes and all the buildings were a similar style. The owner of the estate greeted me, a university professor, with extreme reverence, and later that evening gathered his staff around to hear me share my thoughts about organic agriculture. He had established this estate through a connection with the village leader, who arranged for him to access farmland and hire some of the local farmers as workers on his new organic farm. He did not sell any of his produce but instead used it for himself and his guests. He was fascinated with traditional Chinese culture (particularly Taoism), farming practices, and wild foods. He had acquired an enormous traditional wooden oil press from the nearby village, to make camellia oil. He and his staff ground their own soya milk and tofu in a traditional stone grinder. The farm owner was passionate and had created an idyllic site, but it left me puzzled. How could this kind of operation contribute to or be a model for ecological farming in China? It was so exclusive, only open to one person and his friends. It wasn't even boosting the supply of organic food, given that he sold none of it. When I pressed him on this question of the lessons that his experience offers for others, he had little to say. He wasn't interested in seeing changes in government policy or engaging in public debate about expanding ecological farming and food supply. This farm was one thing that he could do for himself and his network of friends. He sometimes encouraged other wealthy businessmen he knew to try to establish something similar, but that was as far as his advocacy extended.

Some other farms were completely economically driven, with little evidence of concern for ecological dimensions other than complying with particular criteria in order to qualify for selling their produce as organic. At these farms we were sometimes offered boiling-hot tea in flimsy plastic cups, which made me scared to drink from them given the potential for chemicals from the hot plastic to leach into the tea. And at mealtimes we were often provided with a plate, bowl, spoon, and ceramic cup all hermetically sealed in a plastic bundle, a sign of food safety and hygiene in one sense but with seemingly little concern for the waste that was generated in the process.

At yet other sites, we were captivated for hours on end learning about the efforts that buying clubs or restauranteurs and social enterprises had gone to in order to procure wild and ecological foods from farmers in remote areas and present them to urban customers. These alternative food networks and values of food sovereignty (without using this term) were cultivated based on slow but steady labour of establishing personal connections and trust in an arena of paranoia about unsafe food and "scattered farmers" that government bodies seek to standardize and establish surveillance over.

In Hong Kong, I was amazed to experience "Organic Day" festivities, during which time the main thoroughfare in the business district, "Central," on Hong Kong Island was shut down for half a day. There was a breakfast gathering and panel talk including local politicians, and along the streets there were all sorts of family-friendly activities and photo ops to boost consumer awareness and interest in organic food. Very sophisticated marketing.

Zhenzhong Si

I was fortunate to have conducted most of my fieldwork with my supervisor, Dr. Steffanie Scott, and her then doctoral student, Dr. Theresa Schumilas. On our way to farms and meetings, we had extensive discussions about the interviewees and cases we visited. In fact, these fascinating cases seemed to be all that we talked about. This "collision of thoughts" motivated me to reflect on the data we collected and helped me generate new ideas. Being Chinese gave me significant insights in understanding the cultural contexts within which the AFN initiatives have been evolving. Although I tried my best to translate the interviews into English for my colleagues, I worried that my choice of words might distort the original meanings of the informants' responses. But taking notes in Chinese made it possible for me to revise our interview transcripts later to produce a more precise translation.

I had never got to know so many people with such diverse backgrounds before I started the research. The hospitality of the many farmers we visited and interviewed is heartwarming. They invited us to their farms, homes, and vacation houses. I was even invited to live on a farm for three days. They cooked the most delicious dishes for us. They were excited that we were interested in their efforts. With all kinds of imaginations of Western organic farms, they were also extremely eager to learn from us the experience of farmers in Canada. Their overwhelming passion made me feel strongly like a messenger that connected them to an unknown world.

One interesting observation from my fieldwork is that the diverse "nature" of the growers of ecological food in China bestows different attitudes and perceptions on them—about life, health, and the future. Among these growers, some were struggling beginners who were ordinary factory workers with limited resources. Some were established farmers who had lived abroad for a few years and held strong environmental ethics. Some were undergrad students or interns who were still debating with their parents on farming or whether they should go to graduate school. Some were small entrepreneurs who had to abandon organic certification for its high costs. Some were hired workers in organic farms or food companies who used to be conventional farmers and had to convince themselves to take in ecological farming approaches. We also met millionaires who were successful businessmen with extensive networks and abundant capital.

For most of the people, sustaining an organic farm, certified or uncertified, was a challenging task, although the extent of the challenge differed from person to person. Things were normally vast and hazy for beginners and interns. They appeared confused sometimes by the seeming ease of success of other farms. They were eager to seek guidance on farming approaches, managerial problems, and/or marketing strategies. Meanwhile, they were strongly motivated and committed to ecological farming, although they were struggling to remain in business. They did not hold back their delight that we were actually interested in their work and dreams. Interns normally face strong opposition from their family, who believe working on farms contributes nothing to their future and is a dead end. This awkward situation is partly attributable to the modern dreams promoted and imposed on everyone by the Chinese government, which marginalizes and even disqualifies farming, especially if on a small scale, as a component of this modernity. Holding a university degree but working as a farmer disgraces the family in certain cases. This pressure was fatal for their farm career if they could not make ends meet. A young man we met twice in our fieldwork subsequently went on to graduate school because of pressure from his family deterring him from farming. There are many other similar cases among new ecological farmers in China.

Established ecological farmers who were not receiving government subsidies or other support were especially critical about chemical intensive agriculture and the policy environment that supports it. They spent a lot of time maintaining their personal connections with their customers. They worried about their land access in the long run, their unstable customer base, and the rising cost of inputs. Their anxieties were in sharp contrast to the confidence of organic food companies designated by local government as dragon-head enterprises. These companies were like the top students in the class, who received constant care from the government. They treated our visits or interviews as opportunities for publicizing their business, which were normally larger-scale farms with substantial investments. Our visits could be amusing because we were dressed like casual backpackers formally greeted by receptionists in fancy attire, who ushered us into elaborate boardrooms adorned with plaques and galleries of product samples. In some cases, photos were taken throughout our visits and later appeared on their websites. They were mostly interested in branding and expanding. Their relationship with customers was thus more business-oriented and impersonal.

A small number of ecological farmers we met in China were millionaires or billionaires. They were too rich to run their farms as businesses. In fact, they never seemed to care that much whether they made profits. From time to time, their farms reminded me of the classical philosophy *chushi*, pursued by Chinese intellectuals for thousands of years—a constructed utopian land (*shi wai tao yuan* in Chinese) to escape from the chaotic world. Throughout Chinese history, rich businessmen and retired officials have built fancy elegant gardens to enjoy. Intellectuals pursuing alternative lives would live in cottages in mountains. Farming within this context is a particular way of life rather than a business. The several farms we visited interestingly reflected this ideal. Being far away from big cities, they were almost isolated from the outside world. They replicated lives of pre-modern times. One farm established by a successful IT businessman and his wife had a full set of tea houses, reading rooms, and guest houses on the farm. They did not use chemicals in the field or do weeding because of the guiding principle of "natural farming." The very low productivity did not seem to be a concern at all. My interviews with the farmer during my three days' stay on the farm were more like casual conversations amid tea tasting. Everything that was supposed to be taken seriously by a farmer, such as productivity, pest and weed control, soil fertility, and inputs and outputs, was not a problem here. For them, ecological farming was more like a way to enjoy life. On an organic tea farm established by another wealthy businessman, the situation was almost the same.

Theresa Schumilas

I live and farm in a village of about 800 people in rural Ontario. While I was well prepared academically for my first field trip to China, nothing could have prepared me for the contradictions I encountered that truly, as they say, "rocked my world." I share here just few stories and observations that have remained salient with me, even after two years of reflection, processing and further research.

I have been part of the organic movement in Canada since the early days. I "cut my teeth" on books like Rachel Carson's *Silent Spring* and E.F. Schumacher's *Small Is Beautiful*. I read *Farmers of Forty Centuries* for the first time in 1978 when I was trying to figure out how I might get back to the land and make a living on four acres. In my circles, we naively reified China's peasant farmers as the world's experts in closed ecological farming systems. I was so excited at the idea that I could meet and learn from them.

Indeed, I found some of these peasant farmers, but they were not farming. They were either driving taxi cabs or pulling weeds on peri-urban farms operated by young, educated urban residents. While they weren't the peasant farmers I hoped to meet, these CSA operators, farmers' market volunteers, and buying club coordinators were inspirational in different ways. I found all my favourite books on their shelves and indeed built lasting friendships and connections with some of them, and feel a solidarity in our struggle to build more sustainable food systems. They have hurdles that I didn't have to face as a new organic farmer. They have inherited exhausted soil, left bare over too many winters and killed by too many years of over-fertilization and misuse of pesticides. Despite visiting dozens of farms, the only amphibians (which I consider a good indicator of biodiversity on a farm) that I ever saw in China were in the wet markets. Plus, the farmers seemed totally caught in a "productivist paradigm" and looked at me quizzically when I explained that I left part of my farm in a crop that was not harvested for sale every year so I could build soil life. They were also challenged by the absence of agricultural outreach and resources pertaining to ecological approaches. Any such materials that I was able to share were hungrily gathered up and shared in their networks. Here I was, expecting to meet the world's greatest ecological farmers, and finding that I was sharing information about GMOs, seed saving, and basic soil biology.

On the other hand, there are also strong drivers for this grassroots movement in China. Whereas accessing small plots of land to begin peri-urban farms and CSAs is next to impossible for "new farmers" in Canada, in China this is facilitated by common land ownership and lease arrangements. Plus, there is no debate about whether or not these are "real" farmers. In Canada, my four acres often doesn't qualify me as a "farm." But in China, where all farms are small, these CSA farmers are accepted and acknowledged by other farmers, academics, and local state officials. Indeed, in China I was considered a "big farmer," probably for the only time in my life.

I also couldn't believe the number of workers on these small farms. Before I went to China, I had just finished a study of CSAs in Ontario that involved a number of farm visits. At every visit the farmer looked exhausted and rattled off for me their to-do list, and lamented the weeds in the field, the fence that needed repairs, or the tomato pruning that wouldn't get done this year. So imagine my disquiet as I stood in field after field on CSA farms in China and could not identify any task that needed doing. If I could have, I would have jumped in and helped out. But there were no weeds to pull. All the vegetables were cultivated and ready for the next week's harvest. Stalks from the last harvest were in the fermentation pit (their approach to composting). Plants in need of pruning were neatly tied up. Soil was recently watered. All I could think about was how long my to-do list would be if I wanted my farm to look like this when I got back home.

From all of this, in addition to completing my dissertation, I'd say I brought three key "tacit" learnings home with me and integrated them into my life as a woman, academic, and farmer. First, I'm inspired by how possible it is to remain connected with "new farmers" across the world. I had not really understood before how our digital age presents unprecedented opportunities for building networks and global movements for change. This is a new area for research and praxis for me. Second, my understanding of what a "farm" or a "farmer" is has opened up. People in my community talk about farmland being "lost" to sprawl. But this rests on a particular view of "rural" and "urban" as discrete and the assumption that a farm is a big swath of land lying outside of the city. Indeed, if we redefined how we understand farming and farmland, we would understand we have surplus land. The crisis is in how we treat it. Finally, the experience has left me saddened and embarrassed at how silent I have been about environmental degradation and ecological injustice in my own backyard. In China, I watched a group of young (mostly) women crossing back and forth over a shifting boundary of what is permitted in an authoritarian state to press for higher-quality food and stronger ecological protection. Yet, here in a country where the only thing I risk is missing my favourite TV show, I cocoon and do nothing. What has stopped me?

Glossary of key terms in Chinese

sustainable agriculture   *kechixu nongye*

ecological agriculture   *shengtai nongye*

green agriculture   *lüse nongye*

permaculture   *yongxu nongye* or *pumen nongfa*

natural farming   *ziran nongfa*

ecological civilization   *shengtai wenming*

hazard-free food   *wugonghai shipin*

green food   *lüse shipin*

organic food   *youji shipin*

new farmers   *xin nongmin, xin nongren* or *xin nongfu*

urban agriculture   *dushi nongye*

conventional agriculture   *changgui nongye*

industrial agriculture   *gongyehua nongye*

sustainable food system   *kechixu shiwu xitong*

Index

Page numbers in **bold** denote tables, those in *italics* denote figures.

3 20